LAST RANCH

The First Ten Years
1972-1982
POSTSCRIPT

Elizabeth Mann

Order this book online at www.trafford.com/07-0382
or email orders@trafford.com

Most Trafford titles are also available at major online book retailers.

Note for Librarians: A cataloguing record for this book is available from Library and Archives Canada at www.collectionscanada.ca/amicus/index-e.html

Printed in Victoria, BC, Canada.

ISBN: 978-1-4251-1976-8

We at Trafford believe that it is the responsibility of us all, as both individuals and corporations, to make choices that are environmentally and socially sound. You, in turn, are supporting this responsible conduct each time you purchase a Trafford book, or make use of our publishing services. To find out how you are helping, please visit www.trafford.com/responsiblepublishing.html

Our mission is to efficiently provide the world's finest, most comprehensive book publishing service, enabling every author to experience success. To find out how to publish your book, your way, and have it available worldwide, visit us online at www.trafford.com/10510

www.trafford.com

North America & international
toll-free: 1 888 232 4444 (USA & Canada)
phone: 250 383 6864 • fax: 250 383 6804 • email: info@trafford.com

The United Kingdom & Europe
phone: +44 (0)1865 722 113 • local rate: 0845 230 9601
facsimile: +44 (0)1865 722 868 • email: info.uk@trafford.com

10 9 8 7 6 5 4 3

The countryman, better than anyone knows
that in traveling toward an unknown country,
he must continually pause, and look back
to get his bearings, and keep his direction.

R.D. SYMONS

To my husband Henry, and son Eric,
who made this journey with me; to
all farmers and ranchers who have made
a similar journey, this book is
respectfully dedicated.

ACKNOWLEDGEMENTS

I AM DEEPLY INDEBTED to my long remembered helpmates: Arthur McCuddy for his stories of the old days. To George Geldart and Terry Peterson, Regional Economists, for their 1983 guidance and endorsement of the economic information. To Murray Soder, District Agriculturist for his suggestions. To David Holmes-Smith whose clear-headedness and thorough copy reading and editing gave the final impetus to completing this manuscript.

INTRODUCTION

IN THE MID 1800s, with the beginning of British Columbia's population growth, towns began to be built in the most convenient locations in the valleys, on flood plains and alluvial fans where the river had eroded, and then receded to form a broad open plain.

In the same way, early settlers in the interior found that a distance from these towns, a bend in the river or creek often brought forth a similar, though smaller fan, a widening of the valley floor with the much desired tall bunch grasses. There the settler took root, and with his own hands began the rudiments of a ranching operation.

Each ranch has had to form itself nearly as harmoniously as the land itself, as each ranch is dependent upon its own unique conditions: its pockets of land, its water, its cattle movements within the restrictive terrain, its grazing grasses, and its service centre or town. A ranch begins to fail if even one of these conditions deteriorates is altered or removed.

With the passing years, some ranching conditions have been changed. Water is lost, range is removed, then obtained elsewhere, whereupon costly livestock trucking is needed. Broader valley ranches have been carved up into smaller units, or a few select ranchers have expanded to include adjacent smaller units. In the main, however, many of the ranches still remain, stubbornly refusing to part from their affinity to the land.

Traditional practices continue for a Crown range, attached to a specific ranch, to be held by permit or license, and are mutually shared by loggers, fishermen, hunters and tourists. A very small percentage of range is held by long term lease. The additional yearly cost of the lease and taxes entitles the rancher to more control over the land's use, but in most instances, it is also open to judicious public use.

All ranges are carefully monitored by the Ministry of Forests for their grazing conditions and carrying capacity. That capacity is reduced or increased in proportion to grazing conditions. The objective is always twofold: to maintain the grasses for the cheapest and most effective way to produce pounds of beef, and to keep the dry interior grasses in check to prevent the spread of grass and forest fires.

In 1982, in the interior southern half of British Columbia, the ranches and their supporting Crown grazing lands produced upwards of 200,000, or about eight percent, of the country's calf numbers. The prairie provinces native grass lands and pastures produced the remaining million and a half head of calves raised in the west. The eastern provinces produced the balance of the estimated 2,400,000 calves raised yearly in Canada.

Though a small percentage of calves in British Columbia are raised on pasture conditions, high-cost, town-adjacent bottom lands continue to make this uneconomical. Without the support of the nominally-priced Crown ranges, beef production as we know it here could not continue.

However, in the 1990s, circumstances began which could place the future of our ranches in jeopardy. Multinational corporations jostle for massive dairy herds, chicken/egg factories, and large scale cattle feed lots. From farm source to food retail sales, all could soon be owned and price controlled by the multinationals.

If these circumstances persist, our picturesque human settlements amongst the mountainous but often monotonous landscape could soon be gone. Gone could be the long-enduring rancher's choice to produce on land which, in most instances, has no other farming use; gone could be the individual's right to go it alone.

When a ranch is broken up it can never, except in rare circumstances, ever be put back together again. We have tried. This story is our painstaking, often ineffective way to do just that.

CONTENTS

ONE

FOOLS ON THE HILL

The minute you begin to do what you want to do,
it's really a different kind of life.

BUCKMINSTER FULLER

FOR MANY YEARS, even to ourselves, we were to be those fools on the hill. No one locally would consider buying the 800 acres of the deeded property. Their frequent remarks seemed to be:

> *If it was any good as a ranch, it would have stayed a ranch, not become grazing for a valley larger ranch. There's no water there, the creek can dry by August. It's too rough, too hilly, those rocks...*

And even the local high school principal asked, when I registered our son Eric for school, "You're not going to live up there, are you?"

Was it really the outback? Eight miles from town, but within this settled valley, it was considered the hinterland: a place to picnic, shoot gophers, explore, but not a place to stay.

Of course the road was scarcely more than a trail. A trail (in spite of its roughness) that I would learn to delight in through the changing seasons. No telephone or electricity was there to serve us, and the ranch long-deserted, though nostalgic and special to valley residents, was neglected and ravaged.

The house, not more than the size of a cottage, was vandalized and at first glance beyond repair. Windows and doors were broken and hanging, wallpaper tattered and drooping, pack rat and bird nests met your every glance. Broken glass crunched under every footstep. Fools we were to attempt to save it. But as it turned out, we were to have no other choice.

It was obvious that the local ranchers, on observing our inept early efforts, were skeptical. To us they did not say it, but to each other they said, "What the hell are they trying to do up there, lose their shirts? The government should have bought that property for us to use for fall grazing."

Quite soon we were to learn that what the ranchers didn't need were more ranches, or more beef. With an overabundance of cattle and shortage of good Crown range, they didn't need or want our competition. In addition, we had three strikes against us: we were new to ranching, new to the area, and we raised French Charolais cattle in a predominantly long-established Hereford region.

In the valley several good, well-equipped ranches were for sale without takers. It could be questioned why we did not buy on of these going concerns. The truth was that even if we had wanted to, we did not have the capital required (at that time roughly ($1400 per cow unit) to buy an operating 100 head ranch. Secondly, a comforting fact to us was that we could not so visibly make fools of ourselves by jumping that fast into a business that we knew little about; and, thirdly, we did not wish to live within the confines of a mushrooming Okanagan Valley population.

We had carefully chosen this property for the future, had searched for these certain conditions: property bounded on the valley side by an Indian Reserve and circled on all other sides by

Crown land which was protected from development by a logging company's long-term tree farm license. We had found a place in the sun, special and exquisitely beautiful.

It had been a December day in 1971 when we first visited the land. Very early that gray morning, we had driven across the small valley town, and up the winding road, through the rabbit bush, sagebrush and bleached bunchgrass lands to the forested hills beyond.

Scarcely had we reached the trees when the gray clouds—as if whiffed by a tiny breath of air—vanished, and the sky opened to a dazzling blue. Who could have known from the valley floor that the brightness of an Okanagan sunshine beckoned just overhead? But there was the sun nudging us to drive on to those ranch lands we had come to explore.

Everything seemed possible in 1971. A firmly based economy and marginal price increases gave promise to bold new ventures; and our plan, if the land suited us was to find a way to purchase it. With too many years spent in the damp overcast of coastal winters, this unexpected appearance of sun stood as an omen for a fruitful discovery.

As we climbed yet another hill, the last before reaching the ranch lands, the trees parted to an open meadow. Suddenly the sound became empty, birdless, silent, as we looked out, not upon the grassy meadow that we had come to see, but a meadow blanketed in white. Powdered fresh snow softened the rocks, hid the grass, the earth, the soil that we had to be able to see, to feel and to smell. Disappointed and despairing, we knew we had come too late. Further along as we drove the perimeter of the meadow, a solitary blue jay atop a fence post ended the silence, scolding us roughly and mocking us for our tardiness.

It would be spring before we could commit ourselves to the property. It would be late summer before our move could be started.

When we did purchase the land we were dependent on the subdivision of our coast property, then underway, to build a house for the ranch, to buy farm equipment and to carry out the initial land development. We felt reasonably secure that the much-in-demand lots would sell quickly, and that our mortgage on both coast and ranch property could be handled. Within months our complacency was to be shaken.

In December 1972, less than four months following our arrival, the British Columbia government by an Order in Council announced quite unexpectedly the enactment of the Agricultural Land Act, and with it the accompanying land development freeze on all lands containing Class 1 to 4 soil types. Our coast property, consisting of about 50 percent of soil types within these classes, was instantly included within this freeze and the subdivision halted.

The fact that our subdivision was confined to the picturesque rock bluffs of the remaining 50 percent of the land seemed irrelevant. The whole parcel was frozen, still saleable but only as a complete lot.

Following the announcement of the land freeze, the government press releases and news reports (and a letter to us) made it clear that there was "no one to see, phone or appeal to direct." We could not submit to that dictate.

In Victoria we met with Ministry of Highways divisional heads to ascertain their powers to complete the subdivision still in their hands. We met with the Agriculture Ministry to plead our case. Everyone just shook his head, helpless to do anything within the confinements of the land freeze. An appeal directly to the newly established Land Commission was our only route.

Hastily we prepared the appeal. We stressed that the coastal property was not economically viable farm land—we had tried it—that it was easily flooded, that the non-arable part of the parcel was what we wished to have subdivided. We further added that our purchase of ranch land could do more to increase B.C.'s agricultural base, but that our development plans for the land were contingent on the funds to be received from the coast land subdivision, with interested buyers standing in line. Lastly, we protested that the legislation was retroactive, therefore unjust.

We might as well have saved our precious time. The appeal fell on deaf ears. Our only alternative was to immediately place the land for sale as a parcel, to disregard our subdivision and to take our losses. We knew that the price that we would realize would be inadequate for any significant development plans for the ranch. The new house would be the first expenditure to go; all other plans would be analyzed when the funds were received. The longer we delayed listing the property, the worse our situation would become.

It was in the rugged coastal Squamish valley that we had taken our first step away from the pressures of the city and commenced our initial cattle raising. We had chosen Charolais cattle because of their lean meat, and their breed association's extensive bull testing program. As we were neither farm raised nor trained, these livestock evaluations could be a real guide to novices such as ourselves.

While living there, my husband Henry had taken a course in Artificial Insemination and using semen from the best and easiest calving bulls available, had artificially inseminated our few cows.

We had stayed in the Squamish area just long enough to enjoy the valley's unspoiled rain forest beauty, and to realize that it could never be an easy place to raise cattle. With annual precipitation too frequently of nearly a 100 inches, a dry patch of ground was as rare as a breath of wind in that sheltered, humid valley.

Accompanying us in our move was our small herd of 12 cattle, 2 horses, 2 giant-size Irish Wolfhounds, and a perturbed tom cat appropriately called 'Freakie'. Quite soon, we added, 25 fall-calving Charolais bred, purebred Angus cows, and a Charolais polled bull, 'Pollar', a rarity at that time. And before winter, we increased the herd again by adding 24 Hereford heifers which were already bred to Hereford bulls. This was to be the nucleus for our planned 75 female herd. That number corresponded to the near-future carrying capacity of the ranch and its small Crown range, with a net income which would be adequate for our needs..

All that fall Henry hauled the numerous feeders, lumber, containers and our other possessions from the coast. Our small hayfield at the ranch we had custom cut; and by buying the standing hay in a valley field and having that custom cut, we had sufficient feed for our herd's first winter.

With our coast land still unsold, as the summer of 1973 approached, and with it our date for mortgage and cattle loan payments, we knew that some of our cattle would need to be sold to make our payments. The obvious sale had to be our nurtured and cared for Hereford heifers, and, by this time, their fattening calves at side. We saw no other choice.

I recall very well that July morning when the liner arrived to take the cattle away. So easily, the driver backed the long eye-holed metal trailer up to the cattle loading ramp, where in the corrals the cattle stood ready: the grass-fattened heifers, 41 head in number, including their grass-fattened sprightly four-month old calves. The brand inspector quickly checked them off as he stamped his approval to move them out.

We would not be alone in selling breeding animals that year, for many ranchers would be suffering from the year's long drought, the winter's lack of snow and winter kill of alfalfa. Many herds would be depleted by fall to cope with the lack of graze. But our cattle were moving out for a different reason: to pay loans which, except for the protective land freeze, would not be there.

For only a while I watched the bewildered calves, separated now from their mothers, being loaded for the four hour drive. Then I returned to the house, remembering our long nights of vigil and the calving assistance that was required for the majority of those heifers. My body shivered remembering the cold as I held a flashlight through the late February nights, the endless broken nights, sitting by kerosene lamps, checking, looking, and peering every two hours. The persistent questioning, would we save this calf? The delight when we did, the total exhaustion when we did not. But we had failed. We had not lost a single mama heifer, nor had we need of even one cesarian, but seven calves were to be tallied up as losses.

According to the vet, these were not unusual losses in heifer calvings; we should not be discouraged. As was the custom, to obtain smaller calves these heifers should have been bred to Angus bulls, not as they had been to blocky Herefords. Here particularly, our son was perplexed, if not astounded, at the absence of selective breeding by cattle producers. Surely in general, even heifers (if carefully selected and bred) should be able to calve unassisted.

Under the conditions we were forced to deal with, the calving losses had to be blamed on ourselves. Perhaps, in many instances, we had waited too long to offer calving assistance. Without experience or farm neighbours as guides, our only instruction had been a United Grain Grower article on calving. Explicit as it was, with detailed illustrations of normal and abnormal birth presentations and how to correct them, it led us astray in one important detail of calving. It insisted that most cows could do it themselves; that if the presentation was normal, the cow should be left to handle it herself for at least four hours following the initial calving signs. With our *heifers* , not cows, bearing those big, blocky calves, that advice was devastating.

Initially, however, with this article our only guide, we pored over the diagrams, memorizing details like an obstetrician in training. Our retention of the details had to be complete, for our first heifer's calf was about to be born. In the deepest, darkest part of that February night, it was going to be born backwards.

As we roped the fretting mother-to-be, we recalled that we had only two minutes to deliver it once the calf's navel in the birth canal reached the pelvic area. There, the blood supply from the calf's umbilical cord could be cut off from the pressure, and the calf would perish. Only a quick delivery and transfer to its own breathing mechanism could save the calf.

Of course, we did not own a calf puller. Because of an earlier than expected calving start, we had not yet purchased one. Instead, Henry was to build one. In haste, by flashlight, he fashioned from 2 x 4 lumber, a frame in the customary design of a calf puller.

With the use of ropes around the calf's ankles, positioned where the puller chains normally would be placed, Henry pulled gently as the heifer bore down. As he worked, I nervously tried to reassure the frightened heifer. "It's okay, Mama, it's okay." My teeth chattering, I wondered who was more frightened.

In minutes, looking like a rag doll, the small calf slid to the ground. "It's alive. It's alive." To our delight it was very much alive.

Our sprightly beginnings, however, were not to continue. Seven calves later, four had been born dead. Discouraged and at wits' end, I drove to Penticton to discuss our hapless results with an understanding vet.

"Obviously these heifers can't do it alone. Give them an hour; if there is no progress, get out and pull the calf. If you don't, the calf won't likely survive that pressure, or if it does, its head and tongue will be so swollen it won't be able to suck for hours. That may be too late for it."

The vet had advised that we buy a calf puller. Here, with more luck than we had been having, I found the only one in town and of the best type. We felt confident now that our output of live calves would change. It did. Three more calves had been lost, but offsetting that were 17 lively, healthy offspring.

This exhausting initiation into calving taught us one thing. Never would we go through that experience again. Easy calving bulls, and good roomy cows and heifers were essential. And lights, electricity, how we needed bright lights to work by. How we needed hot running water and plumbing; how we needed a bank account to accomplish it all!

In our innocence it had seemed simple to do it this pioneering way; to build a ranch to suit our purebred livestock objectives. We fully believed that we would have the capital to begin it all, but had to recognize now that we would not.

Even those heifers could never have a second chance with us. Not even the splendid few, who would have been chosen to stay in our herd would have a second chance.

Only a cattleman knew the time, the effort, and the energy required for a busy beginner to brand, vaccinate, ear tag and record his animals for performance. We had done this, only to see our efforts wasted. Would a buyer more expedient than we care that this heifer had had problems calving, perhaps always would be too small inside and should be culled? Could he care about the birth weight of the calf, his growth, performance, the mothering ability of the dam? Not likely. The cow-calf pairs would sell on beauty and colour, with the biggest calves and heifers, even if much older, and not inherently better, commanding the best prices. There would be no real concern for intrinsic quality or growth. The useless data could be torn up, if not the experience ever forgotten.

And as the cattle liner shifted into gear, and slowly moved down our rough, bumpy road, I resented governments which dictated Orders in Council, all in the name of protection. Protections of what? Of land. Not us, the farmers of that land.

TWO

IN THE BEGINNING

Why do I judge my day by how much I have accomplished?

HUGH PRATHER.

A RANCHER SO NATURALLY takes for granted those components that are included in his ranch purchase: the 30 or 40 pieces of equipment, the fences, the barns, the sheds and haylands, components that have been gradually added to the ranch, and intermittently renewed. As these integral parts were there when he bought the ranch, he never knew how much time had been spent buying each piece of equipment, building each fence, each barn, each shed or how much time had gone into clearing the haylands.

Painstakingly, we had to experience for ourselves how many hours, weeks, months and years would be given to just this accomplishment. We had to see for ourselves that each day we would encounter endless interruptions by the cattle from our attempted essential improvements.

Because of the loss of our extra subdivision capital from our coast property, initially, everything we did had to be on our own.

And as soon as we saw a good beginning at a mile long cross fence, or seeding a piece of newly-cleared field, or clearing out years of downed alder along the creek, plans had to be abruptly altered by a bull fight taking out a long strip of fence, or by a steer with bloat. Development always had to take second place to cattle needs, and the cattle seemed always to be needing.

Inexperienced as we were in livestock raising, we had to do our learning at a gallop from livestock health manuals and articles. As most comprehensive livestock manuals are written in the United States, we were constantly baffled by which diseases were present there, but not here. No sooner had we confirmed our suspicions that an animal's problems were intestinal worms, or in one case High Mountain Disease, than we were to discover that neither of these conditions was a potential problem in our area.

Emergency situations were always ahead of our studies. Without telephone, we either had to drive to the valley to call the vet and face the 35 mile service charge to bring him to the ranch or do some guess work, diagnosis and treatment on the basis of our careful observation of symptoms.

Even as our background of medical knowledge developed and our experience with reliable medications increased, there seemed always to be blank spots in our total understanding of animal husbandry. Though we quickly learned to recognize the obvious symptoms of a sick animal, the how-to of injections, of administering rumen boluses, of stomach tubing and drenching, we still lacked anything beyond the rudiments of animal physiology. Even the simple moving and handling of animals was, at the beginning, beyond our native understanding. Although we saw quickly the abnormal, we lacked common sense knowledge of the normal workings of a cow to the point that our first uterine prolapse, during our second year, was an experience beyond our knowledge, and even beyond our imagination. It never occurred to us that the complete calf bed could slip out following calving. All we knew when we saw it was that it was a problem beyond us, and in great haste I rushed off for the vet.

This trial and error position which we had placed ourselves in, caused us far greater stress than had we been ranch raised, or had we taken courses in livestock husbandry. Expecting it all to be trouble free, each day we found that our high expectations were regrettably lowered by our experiences. We did not know that across the country, the average calf weaning percentage from cows and heifers exposed to a bull was close to 80 per cent. Consequently, we felt nothing but failure when a calf died or a heifer came home unbred in the fall.

Whenever we could, we gave the time to the required changes to the land. This ranch development worked as a distraction to these early livestock experiences. It was also obvious that the more efficient our layout became, the easier the handling of animals would be. The more and better quality hay we produced, the healthier and happier our animals.

We had given up the possibility of acquiring easier, flatter hayfield acreage. The effects of the land freeze had placed additional land purchases out of our reach. Usable blocks of valley land which had sold for $700 an acre in 1972 were selling for $2500 an acre in 1975. Forested 160 acre parcels had risen from $15,000 to $50,000. Changes in government land policies had priced Crown land at near market value, which put out of our reach large tracts of arable forested land near the ranch. This required that we direct our energies, time and funds to intensifying the production of our own ranch land.

Windfalls and logging debris within even small meadows was piled and burned, and the grasses fertilized for better production and grazing. Systematic, yearly logging was arranged, which provided some of the funds for additional hayfield expansion. Areas logged, but not suitable for hayfield, were broadcast seeded to grasses. Slow but steady intensification seemed feasible and advisable. Grazing for an ever increasing cow herd and larger hayfields to provide winter feed was a necessity.

When we purchased the property we had projected an economically viable herd size of 75 cows. But in 1973 the forced sale of our 24 heifers had pushed that magical figure beyond our immediate reach. By the time we were to attain it, that figure would be as outdated as gas-operated tractors.

In 1973, however, cattle prices were buoyant and the galloping effects of inflation, which began that year, were not recognized as an inhibiting factor. The rise in the price of barbed wire from $16 a roll in 1972 , to $30 a roll in 1974 was regarded as an increase due to a short term shortage—a price increase soon to be corrected. That these high prices were to be the basis of further price hikes did not cross our minds.

The summer we arrived at the property, we purchased our first piece of farming equipment, a very old Fordson gas-powered tractor. We had eyed the tractor at a dealer's lot, priced at a low $200. We saw it as a possible beginning tractor for the ranch; one we could afford. Not sure that my husband had ever driven a tractor, I was relieved when he said that our mechanic friend Omer would check it out. What we didn't know was that Omer, in his cocky manner, was approaching the dealer with a bargainer's certainty. Following a careful check of the tractor, and a quick run around the lot, Omer indignantly said to the salesman, "I don't think I can advise Henry to offer more than $25 for this tractor." The salesman shrugged his shoulders and replied, "Well, if that's what you think it's worth, I guess I can live with that."

Thus we had a tractor, and for our first year that was to be our status as farmers, that and a set of diamond spiked harrows.

The stiff monster of a tractor was not as easy to use as we had anticipated. For a few dollars for parts, and the free labour of our thoughtful friend Omer, it was made fully operational, but slow, long and awkward, it proved to be totally useless for haying. Again, the second summer we had to have our hay custom cut.

In the long run, however, inflation and a 1973 shortage of farm tractors did work for us. Locating a solid 1969 Ford 4000 diesel tractor the following spring, Henry offered the old Fordson as a trade.

"Does it run?"

"Yes, not much power, but it runs."

"Well," the dealer pondered, "guess against this Ford I can offer you $1000 for it." Henry, holding back his excitement and surprise, quite rightly made a deal.

Now we had a real tractor, with a loader soon added, and with this, a sickle mower, a side delivery rake, and a Massey Ferguson baler. That summer Henry hayed for the first time. That year we continued to clean out fields, collected fire wood, piled and burned nearly a mile of downed alders and willows along the hay field creek, repaired fences and cross fenced. On the house, Henry had time to finish the plumbing and with a battery operated toilet installed, as a crowning achievement, in ceremony, we burned down the leaning tower of Pisa, the outhouse.

A 16 month drought from the fall of our arrival to 1974 convinced us that irrigation had to be an essential part of our development. We had listened to local ranchers protest the need for it. "Hell, it's like a milk cow, you're tied to it like a baby to its mother." But their circumstances were different; they had a ranch; they had endured and paid it off. We could not afford to be old fashioned; survival for us depended on our being a step ahead.

As with everything we did to develop the ranch, the commencement of electricity and the irrigation system was permeated by unforeseen difficulties.

Our massive diesel plant, an old but sturdy 5000 watt Armstrong Sidley, had sat idle for over a year while Henry strove for time to build a generator building, then to find time to learn how to wire the house, the pump, the generator building, then to have the quiet time to do it all.

If I remembered nothing of my school studies of electricity, let alone knew the mysteries of capacitors and rheostats, voltage regulators and generator brushes, Henry had chosen to remember little more. But lack of money, his determination, reading and asking the right questions had refreshed his memory enough to do the electrical job. But never would it be as simple as the flick of a wand, or as convenient as contacting a hydro office to connect up a line.

Determined to have electricity for our second winter on the property, all was ready for a test run late September. The motor was cranked and started, a switch thrown, and we watched as the voltage meter began to climb, but to our chagrin stopped at 10 volts, resisting even an attempt at reaching the required 110 volts. Henry checked this meter, the wires, the connections, then threw up his hands and through a groan said, "The plant must be faulty."

"But it was working when we bought it," and as I said the words, the truth of the situation struck me. "You mean we have to pull it out of here for repairs?"

"Looks that way; at least the generator."

Days later we backed our only vehicle, a '68 Land Rover up the slope to the generator building. I hesitated, wondering how we were going to lift that 500 pound generator up and out of the door into the enclosed back of the vehicle. City bred as I was, it was usual for me to see anything physically demanding as nearly insurmountable, when in actuality, doing it was really very simple. Boards were placed from the tail gate of the vehicle down to the generator in the building, then, with ropes, the round-bellied beast was guided and tugged, and slid along the boards into the vehicle. It was loaded in minutes.

As the 'gofer', the main road-runner, I was volunteered to drive the generator to the nearest electric motor repair centre in Penticton, 35 miles away.

I drove into the city, our major service centre, that I had not even had time to explore, and a last located the repair shop, and at the rear, the unloading ramp. Inching the vehicle carefully up to the ramp I felt as if I had moved the generator a thousand miles. Forgetting momentarily my knowledge of all cities, I felt like a pioneer of the Arctic tundra rolling into a northern repair centre with a precious and essential piece of technical equipment.

I was astonished to learn that the receiving clerk regarded the generator as just another vacuum cleaner in for repair. My protestations to the clerk led nowhere.

"But they need their vacuum cleaners too," he insisted. Finally, exasperated with my inability to communicate to him our desperate life-death struggle for survival, I impatiently quipped, "Well, do your best to get on it fast. We don't have a telephone; I'll call you in a week."

It was to be another month before the generator would be hauled back to the ranch and re-attached to the diesel motor. Worn brushes and corroded commutator rings had been the reason for the power loss.

Finally, the day arrived for a re-test of the plant. In excited anticipation, I turned on all the lights, and Henry repeated the earlier operating procedure. But this time, Voila! Lights, purer and brighter than hydro could ever provide. We jumped for joy, feeling as excited as an Edison; then I rushed to do the vacuuming. How rare a treat vacuuming can be, when a broom has been the only cleaning tool for over a year.

Our first irrigation venture was to make use of the creek, supplemented by an adjacent shallow well. This system was operated by a 2 h.p. electric pump, energized by our diesel plant.

It required two persons to start the system: one in the diesel plant building to crank the plant motor for the start, and to turn the rheostat up to provide the necessary power to start the hay-field pump; the other in the pumphouse to ensure water flow, with hand signals the only means of communication between the positions a thousand feet apart. Once a pump was operational, the rheostat was turned back to the normal setting, but to regulate the load, even in summer, we had to have house lights on to utilize the excess power generated. If the pump was functioning, this was not as essential a requirement, but if the pump should fail or be put off for sprinkler setting changes, the house lights, or an outside heater had to be on to balance the load.

During those two summers of operation, life in the house was, at any time, jarred by the yell of Henry, or the arrival of Eric at the house, his voice not yet loud enough to be heard, "Turn on the lights, quick!" It was not my place to question why; I did what I was asked, ever confused as to what was happening.

Powered by this diesel plant, our simple irrigation system provided us with about 15 acres of irrigated hayland, half of which we had reseeded to alfalfa. The 25 acres of the original hayfield remained unirrigated producing only one cut of hay. The total hay yield did provide just enough hay for our 40-cow herd, but as we were already learning, that number on this land, could not begin to pay their operating costs plus the basic improvements needed to handle them.

By the time our receipts from the sale of our coast property were received in early 1974, mortgage interest and debts had devoured all but half of what we projected we would receive had the land been subdivided. The balance of these funds had to be used for basic farming equipment, not the proposed land and structural development, nor a new house, as had been planned.

Now more fully aware of the need to supplement our inadequate income, Henry learned of a snow plowing contract available on the ten mile stretch beyond the ranch to a microwave section. He was assured of the contract if he had the necessary snow plowing equipment, a bulldozer. With additional reasons for owning a bulldozer, he began in earnest to locate one that we could afford to buy and maintain.

Every rancher can use a bulldozer. Even I could see the advantages of being able to push brush together, to clear and pile snow from our roads and calving areas. Scarcely had I thought through the advantages and disadvantages that loomed brightly, when Henry brought home news of a local bulldozer for sale.

"What do you think they will want for it?" I questioned.

"It's supposed to be old, a '49 Allis Chalmers, but well maintained. Hughie, who knows the owner, thought about $500, but that seems a bit low." was his answer.

The next day Henry took along a well-qualified heavy duty mechanic to examine it carefully. The mechanic pointed out that the motor seemed very sound, the tracks and undercarriage were in about 75 percent condition.

"It's limited in its use, but it should do a snow plowing job and serve your needs."

"What do you think I should be prepared to pay?"

"Oh, I don't know, it's old, but these old machines go on forever; $2500 would be a good price."

Later, Henry was to approach the owner, a large well-established field crop grower. Remembering what Hughie had said, and not what the mechanic had suggested as a price, he boldly opened the bidding at a lower amount.

"Hughie said you might want $500 for it, but I can only make an offer of $300."

Expecting to see a stunned look from the owner, he was dumbfounded to hear the owner say, "Well, I don't need it anymore. I picked up a better one for my needs; it's yours."

Thenceforth was to begin my involvement in one of our most trying winter ranch jobs: the snow plowing contract. As usual, my role in the endeavor was never adequately explained until the contract was won.

This snow plowing from the ranch, to the microwave station was contracted on a B.C. Telephone demand basis. Usually several plows were done each season, necessitating, not as I had imagined, an up-and-down plow in one day, but often, during those heavy snowfall winters, four to five days of approximately six hours a day of actual work.

The demanding routine during those plow days was for me to drive our son to school. In the interim, Henry would feed our herd, then begin his plowing by making one or two passes over possibly three or four miles the first day. Late afternoon I would pick our son up at school and deposit him safely at home, then proceed up the narrow, high-banked snow tunnel of a road to locate Henry. I would carry extra diesel oil, a kerosene heater and tarp for the cat. By about 5:30 p.m., we would be back down at the ranch, and I would commence my evening meal preparations.

The next day would proceed the same, except that I would be needed to accompany Henry up the road to where the bulldozer was parked to see him safely away, then to proceed with the Land Rover for home. Late afternoon, after returning our son from school, I would again drive the dark tunnel of snow, this time over more miles, to locate Henry.

During these winters, the chance of meeting another vehicle was not likely. This consoled me, but did not reassure me, for there was always the apprehensive feeling of possibly becoming stuck, or not making it up one of the steeply-pitched turns in the road. Though I knew in principle the way to put on chains, I doubted that I could handle them with the light from a flashlight and within the confinements of the walls of snowbank.

Each trip up, my mind raced with the thought of my stranded husband many miles above, as he paced, waiting beside a cold, darkened bulldozer. I urged the gutsy Land Rover on, but often as I peered through falling snow, I cursed the designers of those silly, inadequate windshield wipers that rain-torn Britishers attach to their Land Rovers. With no walkie-talkies or CB communication and the nearest telephone too many miles and hours away, I silently wept for the shelter of a city home and shivered from the awareness of my aloneness and vulnerability.

But except for the curse of my endless fear, the toughly built Allis Chalmers cat did its job, my husband his, and I mine. But with the sight of March's last snowfall and the raven's first call of spring, I felt the release from a burden beyond myself.

THREE

ALARM CLOCK

An appreciation of the process, makes uncertainty bearable.

M. FERGUSON

JERKED INTO THE blackness of another 2 a.m. awakening, I twisted and wrestled until I found and swatted that source of mortifying sound, the alarm clock. This was the first of more than two months of cruel beginnings during this year's calving. It was the same during last year's calving, and would be the same next year.

Within the light cast from a large flashlight, I opened an eye to watch my husband slowly, heavily pull on his grey wool Stanfield underwear, to watch as he sat on the bed pulling on his double layer of four-pound socks, his jeans and shirt; to watch as he picked up his flashlight and moved through the door, and down the stairs to his jacket and hat, his felt-lined boots and gloves. I closed my eyes, having learned not to open them again until he either called, needing my assistance, or the cold of his body slipped back into bed beside mine, aware as I was of the luxury of the warmth he would experience when his body touched mine. It might be twenty minutes before his return, or as long as an hour, but he would return, perhaps only to reset the alarm for 3 a.m., or 4 a.m., only to go through the same procedure again.

Some ranchers avoided alarm clocks by spending these critical first calving weeks in a living room chair, dozing and awakening, checking cattle and slumping back in the chair for another exhausted nap. This way they avoided the soul cringing alarm, but risked in their chronically tired state sleeping too deeply, awakening at dawn to rush out to the sight of a dead calf, or worse yet, a dead mama heifer and her unborn calf. Ranching is that way. The very night you sleep through is the night that the heifer, who can't do it herself tries and fails.

Other ranchers reduce their risks by moving their suspected imminent calvers each evening into barns, but in some instances leaving outside in error and exposed to the February cold, the very ones that calve. The long seasoned rancher has a skilled eye in judging calving readiness. The critical decision is based on a softening and dropping of the muscles surrounding the tail head, but even in this judgement he can be fooled.

Henry chose to sleep his short sleeps in comfort, tolerating as I did the alarm clock. Later, however, with electricity at the ranch, we rid ourselves for all time of the cruel sound of the alarm, replacing it with the often as disturbing noise from an all night disc jockey on a radio alarm.

Remaining are those ranchers who simplify it all; they calve in April or May when the morning sun comes earlier and warmer, and the winter cold is no longer the newborn calf's greatest enemy. Enjoying as I did uninterrupted sleep, I wished we too could calve in later spring, instead of beginning as we did in early February.

Reasons for calving early are numerous: the calves are larger for range turnout in May, and by fall, the heavier weight calves bring in more dollars from the traditional fall calf sales. With calving primarily out of the way by April 15, time can be made available for farming the land, harrowing and fertilizing the field, discing and re-seeding. Time can also be made available for the indispensable fence repairs before turnout.

Later, for us there was an additional reason for early calving. With our herd on performance testing and with off-spring registered and recorded, sire identification of calves was essential. The cows, therefore, had to be bred in one of the seven single-bull spring pastures before being turned out to the multi-bull summer range. Late calvers, if worthy of registration, required expensive blood testing for sire identification. This we preferred to avoid.

It seemed easy to question why more research was not done on synchronized calving. Though some use has been made of protaglandin, we abandoned its use after one unsatisfactory attempt. The unfortunate result, for us was that the breeding of our test group of cows was delayed a complete cycle. We could not afford to experiment further.

It is interesting to note that it was a rancher and not a researcher who, by careful observation, came up with a plan of feeding to avoid night calvings. Gus Koenefal, a Manitoba Hereford breeder, observed that evening feeding tended to occupy the cows through the night, therefore discouraged calving. We turned to the Konefal Method and saw our expected 65 percent night calvers drop to 25 percent. The night cattle checks were not completely eliminated, but it is far more pleasant to deal with problems by warmed daylight than by cold night darkness.

A calf born in the cold of night, unless up and licked dry by a concerned mama cow, strengthened and nourished by the cow's first milk, or dried with warm towels, or a heat lamp by a concerned human's intervention, can perish within the hour from exposure. Its line to life is often dependent on the mothering ability of the cow. Heifers, quite naturally, confused by their first calving experience, have the least developed mothering ability. Angus cows traditionally are known to have superior mothering ability with the cows persistently and steadily rolling and prodding the calves to their feet.

Our Hereford cows seemed more keen on passing and devouring their afterbirth, erasing the scent from predators, rather than tending their newborn; while the young Hereford heifers were as likely to drop their calves, then return absentmindedly to their place in the young teenage coterie. Often too, a heifer calving in the midst of other heifers can be ganged up on by their curious sisters and be pushed away from her offspring; this causes an imperfect imprinting for heifer and her calf.

These young calfers we had to watch the closest. We soon learned within a ranch situation the importance of mothering ability, and scored heifers for the presence or absence of this trait. We culled ruthlessly for calf abandonment or rejection. Heifers who abandoned temporarily, but brought home in the fall a high performing calf were given a second chance. Usually a second calving brought the heifer around; in some instances it did not. The heifer losing her second chance was put on the sell list, and as frequently her calf, for the calf suffering battered syndrome is as likely to continue the same behaviour with her newborn.

The mothering-up of heifers (calvers for the first time) with their calf can be a frustrating, time-consuming part of a rancher's life. Containing the heifer to force her to give milk to her calf, or milking out and bottle feeding, or stomach tubing the calf can rob the day of precious hours. But with the importance of the dam's first colostrum milk, rich in antibodies, this cannot be neglected. Some large ranches are known to milk out each heifer and stomach-tube the calf rather than risk the calf's not immediately receiving this critical protection. With most ranchers calving out at least ten percent of their female numbers as replacement heifers, this is surely not the way to go.

Within a calving season there can exist situations that require interference: a heifer or a cow's death with the calf surviving. This demands an effort to graft the orphaned calf to another cow who recently lost her calf. Some ranchers skin out the dead calf and blanket the orphan or *bummer,* as it is often called, with the hide, in the hopes that the replacement mama will be settled by the scent sufficiently to accept the orphan. This is a hit and miss attempt, dependent on time and patience, with no easy prediction of outcome; yet the struggle goes on to make a profitable union from a loss situation.

The odd rancher even has on hand *bummer* replacements; in this case purchased bull calves from the dairy sector, for just such a calf-loss situation. According to most vets, this is a guaranteed way of bringing new strains of disease to the herd and can never be recommended.

Calving is the keystone of the whole ranching business. It also involves the highest gambling odds. The way the calves are to drop will be the proof of one's breeding, feeding and culling the year before. Bull semen testing in spring would confirm or deny the health and vitality of the bulls being used. This would be no total guarantee of continued bull virility, but it will reduce the odds. November pregnancy testing would provide the initial conception rate, and some guidance on how tight the calving would be. How abundant and green the grass was; how well the bulls circulated; how quickly the cows had returned to breeding condition following their previous year's calving would all be evident.

On a range breeding program we seem to have to accept about a 95 percent conception rate. For a variety of reasons there would be open animals: loners who had avoided a bull, heading off with their calf far out of reach; those with non apparent uterine infections which inhibited cycling, the odd cow with a relish for pine needles, or other toxic plants, aborting from the stress of the toxins; and the odd heavy milker giving out too much to her calf to re-breed under range conditions. The most difficult group to re-breed were the two year olds, particularly those growthy heifers nursing a calf and continuing to grow at the same time. Without adequate lush green pasture or range, or grain and vitamin-mineral supplements, these heifers could return in the fall unbred.

The odds on successfully calving are also influenced by the

health of the cow through the winter, her age, the quality of the hay harvested the previous summer, and used to sustain her through the long winter. These are important factors to her satisfactorily paying her own way. Many a well-fed cow has lost her calf not from the lack of a full belly, but from the impoverished mineral and protein content of the feed supposedly sustaining her. Many a cow has aborted from the eating of molding hay camouflaged in the heart of a bale, and through a severe winter an old cow has a more likely chance of sacrificing the calf growing inside her in order to spare her own life. Lastly, there are the accidental reasons for abortions, or falls on icy feeding areas which can also make a claim on the unborn calf. All this we were to learn slowly, painfully during our early years on the land.

Without the use of vaccines against disease such as Vibriosis, Infectious Bovine Rhinotracheitis, and Bovine Virus Diarrhea (BVD), abortions could be much higher. Like many beginning ranchers, we had to learn the hard way about the need for these vaccines.

During our fourth calving season we made a purchase of several young heifers. Even though these heifers were separated by two fences from our calving herd, we experienced an outbreak of BVD, a high fever, diarrhea condition which caused abortions in susceptible animals. This virus was to cost us nine abortions in our two and three year olds. Initially, the only indication of anything amiss was a slight runny nose in one of the purchased heifers. But within a week the virus had mutated and multiplied within our isolated and thus unexposed herd. It infected at least ten of the steers in the adjacent pen and before we could have the disease identified, it had jumped across the road, literally carried there by boots or tractor, to infect our calving herd, specifically the previously unexposed younger calvers.

Ten days of exhaustive care, expensive time, medications and veterinary investigations and it was all over. Here we gave thanks for the support of the provincial veterinary staff, particularly a Dr. McIntosh, who morally carried us through this trying time.

To avoid unnecessary risks and extra feeding costs, ranchers must be ruthless in each November's cow culling. For us this meant not culling every open cow, but nearly every open cow. Heifers and young cows, providing their calf's performance was in the top 25 percent and providing their breeding period was limited by a late calving, occasionally were given a second chance to grow and strengthen from a free riding year. Usually, the extra growth through that year, and the performance from her subsequent calves amply rewarded such retention. This leniency, however, is rejected by many established commercial ranchers whose primary objective is based on a calf from every female wintered. They rightly argue that retaining freeloaders is a route to building herd infertility, and the objective is to produce a cow adapted to its own range conditions. Therefore, if the cow cannot perform, cull her. Unfortunately, range conditions vary from year to year and in following this line of thought, one is often culling off his top performing cows. The rancher may, in fact, go into the winter with every female in calf, but the calf weaning weights may begin to suffer.

Our pregnant herd was scrutinized and culled more vigorously than most cattlemen would cull. Cows and heifer replacements, whether purebred, percentage, or grade stock that were showing feet, leg, or udder problems were culled. Visible signs of vaginal prolapse were culled as well as that female's offspring. We took no chances with breeding this weakness into our herd.

As we always strove for a more productive, growth-performance herd, in addition to these physically informed culls, the bottom 15 percent of our bred females was routinely shipped, and replaced by our best performing yearling heifers.

During the few years of B.C.'s Beef Income Assurance program, many ranchers thought we were spendthrift fools giving away the potential for extra government handouts represented by those unborn calves. Our long term herd objectives had to take precedence over an easy subsidized route. For us there was no other way to ranch.

Our ruthless dedication did pay off; and it was ruthless and often depressing to send off a favoured cow who had adequately rewarded us each year for our care of her. Then to ship her, perhaps because of a foot or leg problem, not always of her making, but the result of the rough terrain through which she had to travel. But these were the conditions we had to work within; we had to be guided by them, and after the initial years we benefited by having a 90 to 95 per cent weaning in a herd containing up to 35 per cent first calf heifers. But even with this record, one cannot ever be complacent.

Once the calf is on the ground there are a dozen diseases ready to claim its life. Every producer must quickly learn how to identify and treat the most common calf diseases of all: the varied strains of diarrhea, commonly known as *scours*, and an equal variety of pneumonias. During wet calving periods, without exhaustive attention to these stress-related illnesses, a producer can suffer severe and instant calf losses. Unfortunately, even with care and attention, there is no guarantee he will always succeed.

Even with diligence, the building of a super herd is long-time process. Whenever we became confident that we had developed a few lines of solid performers—easy keeping cows with good udders and feet, feminine and fertile with the required mothering ability, but not aggressively protective against us; —we suddenly were presented with an ugly duckling. We could, of course, argue that on the range, the calf may have been sick or separated from its dam. Sometimes it was just that, but it also could be a throwback to a combination of genes, long hidden.

The breeding of solid bulls can be equally as disappointing with the highest performing, most presentable bull, often losing his rightful sire role by inheriting imperfect feet or a low quality semen. Instead, he had his life shortened by going to slaughter.

From the very beginning of our experience in cattle raising, we learned how imperfect, like humans, the bovine is. Sensing, at first that it might be just our herd, we soon learned that in most herds the super cow is rare, or sadly missing, and that the super herd is nearly nonexistent. The imperfect herd and its inherent problems may be the romance for ranching, for many cowboys, but for us economics and our nature dictated that we move in the direction of the super herd.

There are, or course, good reasons for the near absence of these herds. The reasons centre primarily around the pathetic economics of livestock raising, and its cyclical pattern of high and low prices, which cause instability. During times of depressed prices, many good cattle, as well as cull cattle are sent to slaughter. But during buoyant price years the cattlemen attempt to recoup their losses by keeping any cow which might produce a calf, and hold back as herd builders the heifers which should be fed and slaughtered.

Furthermore, during depressed prices, many cattle producers tend to turn to one of their own home grown bulls for breeding, unable to provide the funds to purchase a top performing bull from outside. These approaches only lead to a weakening of the herd quality.

Breeders too are often guilty of buckling under to the same cyclical thinking. To make a dollar can be, often has to be, the main motive of many breeders, who cull only when demand is not there. Bulls and replacements that should have been slaughtered end up in herds, supposedly as herd improvers, but in fact do nothing in that direction.

It seems true that the incentive for super class herd building can only exist within a market stability and price, which includes the producer's cost of production, and the government's adequate funding of livestock research. To date, any recent research we have seen has been performed on too small a sampling to be conclusive. If there are or have been significant research findings, the artery for dissemination of this information is clogged. The rancher is the last to know.

Unlike the highly productive, genetically engineered chicken, there seems such a long way yet to go to identify, and produce something as elementary, as a good, but safe bull to breed heifers. The beef herd represents the last frontier for meaningful genetic livestock research. When will it start?

FOUR

PROLIFERATIONS

Teach us to care and not to care even among these rocks.

T. S. ELIOT

DURING THESE EARLY years we were never pleased with our calf crop. Newborn calves were often limp and slow, a fact which required our constant vigilance. Too many of our potentially growthiest calves were born dead. Not through fault of the heifers or cows; they did a splendid job of calving without assistance, not through fault of the bulls; they did a keen job of breeding the females. The fault here had to be with ourselves.

Very quickly we knew what the problem was: a selenium deficiency in our range grass and hay supplies.

Selenium is a trace mineral essential to the development of a healthy heart and muscles, but the soils of many areas in the west are deficient. Ours was one of them.

Scientific knowledge at that time seemed to concentrate more on the toxic affects of high selenium levels. Not yet able to identify the safe levels, the Health of Animals Branch made selenium nearly impossible to obtain.

There were only two solutions to a selenium problem. First, to feed oats or bran which should have reasonable levels of selenium. But, in a ranch situation there was no guarantee that each and every cow would obtain her share of the grain. Second, to use selenium selenite which was only available through veterinarians, to be poured at a prescribed rate on the hay supplies. This seemed a totally impracticable solution.

By 1975, injectible selenium was on the market. A recommended dose was one c.c. per calf at birth. This we routinely gave, but saw improvements only in those calves with slight weakness.

Throwing caution to the wind, we decided to experiment with one big bull calf whom we tagged with the name of 'Junior'. The calf at a day old was still unable to stand. Instead of one c.c. of selenium, we gave him two, and a day later we gave him another one c.c. For a week, each night Henry or Eric would carry Junior to the house where he had the use and warmth of our back hall. Each day Henry carried him back to his dam, supporting the calf four times a day for him to be nourished by his dam's milk. The fourth day we gave him another two c.c. injection of selenium. What did we have to lose? We know that the calf on its own could not survive. By the fifth day Junior had not died, in fact he was standing unsupported. By the sixth day he was on his own, growing stronger and daily more vigorous. Within ten days he was a normal healthy calf, and survived to grow to his prime.

Even though this cure gave us some evidence that the selenium problem could be solved this way, it was inconceivable to handle each and every weak calf this specially. There had to be another direction.

Consultations with our veterinarian convinced us to try the costly and awkward method of injecting the pregnant cows prior to calving. Since the selenium retention in the tissues of the female was approximately two months, this injection had to be done when they were big, fat and pregnant, not an advisable time to move females through chutes. We did follow this routine, however, for several years; and except for the occasional weak calf in the late calvers, our weak calf syndrome seemed resolved.

Finally in 1979 the Health of Animals Branch cautiously allowed selenium to be placed in cattle mineral mixes. After that, by using an excellent, palatable mix, we completely rid ourselves of this condition.

We were, however, never able to keep our minds centered for long on just the cattle or the ranch development.

It was late one February afternoon in 1975 that Henry startled me by saying, "Do you know what I found out at Northwood today?"

He had gone to the logging company to discuss dam development on their deeded land above the ranch. Interested in news of the outside world, I lifted my head from the screwdriver to say, "Tell me."

"B.C. Hydro has their new transmission line scheduled to cross our hayfield with a station slated right in the centre of it. In fact, they have designed it to cross every piece of deeded property up here."

With over a 100 square miles of Crown land, north and east of these deeded lands, the plan seemed incomprehensible.

"I can't believe such stupidity," I replied, as dumbfounded by the news as he had been. "There must be some mistake."

Within a few days, Carleton, a naturalist who had a summer place above us, dropped by to hand us the B.C. Hydro Feasibility Study for the hydro line. After reading this, we recognized that the engineers had investigated the line's impact on ecology, wildlife, forestry, terrain, but ignored its impact on the people who lived on properties under the designated route.

Swiftly we went into action. There would be no 500 K.V. transmission line through our properties without a strong, determined fight from the property owners affected.

Letters of objection were sent to the engineers of the Feasibility Study, to our M.L.A., the Land Commission, the Ministry of

Agriculture and B.C. Hydro to oppose the location of the line. We searched for adequate reasons for protest: the historical past of our ranch, the number of private properties the line would pass over, the tourist attraction in the hills above us. But, most importantly, our main argument was that they could easily run the line through Crown land north of us, a *shorter,* more direct route.

Before the spring was over, we met with officials from B.C. Hydro and the engineers of the Feasibility Study. Our arguments were adamant and to the point. In fact, as a group, we could have been criticized for over-kill. Nevertheless, with B.C. Hydro's previous reputation for often unjust land dealings, we could take no chances. What we did not know was that a new cooperative image was in the making by B.C. Hydro. They were eager and willing to look at alternative routes.

By early summer their decision was sent to us. In jubilation we read that the line would, and could easily, be relocated a mile to the north of us. In retrospect, we wondered why they had wasted our time.

In the midst of all these proceedings, Henry returned to talk with Farm Credit Corp., with the very same loan advisor who had in 1972 scoffed at our farming and ranching intentions for the property. He had, at that time, attempted to send us to the Peace River to take over a re-possessed property there. We had laughed at the idea, insisting that we were not tough enough nor experienced enough to handle unknown problems there in the north. He had, as humorously, rejected financing our purchase of the ranch property. We had been refused, but not disheartened. We had parted to obtain our mortgage through the local Credit Union.

But return to him Henry did. We could not continue to carry that increased to 12% mortgage, and if the Farm Credit advisors could now be convinced that we were credible farmers, we might be able to obtain a lower interest.

Henry had prepared a lengthy, detailed plan of development including the additional clearing of 70 acres, then underway; the purchase of more cows, better haying equipment, and irrigation development for the 110 acres. This would require that electricity be brought in from its terminal four miles from the ranch, and the construction of two irrigation pipelines: one from our McCuddy Creek, the other from the neighbouring Baldy Creek, over 1 3/4 miles away. A small dam on a natural kettle was to be built to store 27 acre feet of water for late summer use. This Heart Pond reservoir would be connected to the pipelines so that water could be used either from the McCuddy Creek source early in the season, or from the reservoir later in the season. Water, gravity fed from Baldy Creek would irrigate the upper fields.

Henry presented this plan to the Farm Credit loan advisor.

"So in spite of my advice, you bought that property. Well, I'll come out and have a look at it," was his amused response to Henry's proposal.

He did come out and was suitably impressed by our three years of full-time dedication, and by Henry's further plans for development. Though we could not call the land yet a ranch, evidence at least indicated that the property was under development, that someone again lived there. The Farm Credit advisor, George, a crusty, humorous fellow, and a man for all seasons, was from that day forward ready to pay attention to us. We would talk again and again. He would often throw up his hands at our madness, but he would never again fail to take us seriously or to back our expansion mortgages.

Thus in the following spring, with a mortgage double our original, we launched into the second stage of development. Additional bank funding had to be obtained for equipment and cattle purchases, placing more pressure on us from these suddenly rising five-year prime plus one percent loans, than if Farm Credit included these purchases in our mortgages. However, with the use of a Hesston stacker, haying became a one man operation, Henry's, and by late summer 1975, our cow herd numbered 75, that magical number at last!

We knew that the Heart Pond reservoir was a short-term compromise. Henry had done extensive investigations of the adjacent watershed; he knew that water storage was a necessity and was feasible within our immediate area. With this in mind, he had expanded our original water licenses to include larger water storage provisions.

Originally, a feasible, larger dam site had been located on our creek within 500 feet of the ranch headquarters. With the use of his engineering knowledge, Henry surveyed and designed a dam and forwarded these drawings to Victoria and obtained design construction approval.

The logging company removed the standing timber from the seven-acre site of Crown land where the reservoir would be located. Soil tests had been done for us privately; as well the government Water Rights Branch had taken samples for their own tests. We were given every reason to believe that the adjacent soil materials would be suitable for dam construction.

This dam and reservoir had been planned for our second stage of development. With only the construction contractors to be hired for the job to commence, the Water Rights Branch again returned to do a second series of soil tests on the available, adjacent dam-building materials. They had, it seems, lost their original samples.

Nearly at the zero hour of construction we received a terse letter from them, a letter which was to crush two years of efforts and our spirits. The letter read:

"The material does not contain enough clay. I suggest that you find another borrow pit on your property."

With proximate materials deemed unsuitable, our only alternative for a dam at the planned McCuddy position was to locate a borrow pit elsewhere, and haul the material back to the site. The costs of this would be immediately prohibitive. Thus abandoning this, the Heart pond reservoir, though smaller, was chosen as a less costly expedient alternative.

The pipelines serving this were modestly simple. It sent water down from the creek above, under such intense pressure that the force drove the water back up a hill and into the Heart Pond reservoir to fill it. This pipeline was also used to carry irrigation water part of the distance to the hayfields.

The two miles of heavy plastic PVC mainlines to the Baldy Creek—from an eight inch size, reduced down to four at the ranch gate—were laid over rough, rocky terrain. A courageous bulldozer operator was to dig a trench over steep hillsides for line burial to a depth sufficient to protect the line from falling rocks or trees and from sun damage. With the use of dedicated local men and our hired man Angus, the lines were laid quickly and the construction of them was relatively trouble-free.

By late summer of 1976 both lines, the Heart Pond-MCuddy system, and the Baldy Creek system, were operational. Irrigation was underway on our total 40 acre original hayfield, and an additional 10 acres readied and seeded from the new clearings.

To develop more land for irrigation was a more trying undertaking. Root and rock picking on the logged and cleared slopes was to be laborious work for all of us during the following six years. It was mainly our son, Eric who pushed on with the help of a hired hand or summer student assistance. Some acres were picked with the help of local groups such as the square dance club or the teenage club; none could be picked with rock picking equipment.

On the hillsides this equipment seemed useless; the combination of rocks and roots jammed up the machines. We had to fall back on the tractor loader, a trailer, and, later, an old dump truck.

Only in the latter years was a root rake located. This proved time and back saving, by windrowing broken roots and branches for easy pickup by the tractor loader.

FIVE

RANCH HISTORY

There is no liberty except the liberty
of someone making his way toward something

ST. EXUPERY

ONLY A HUNDRED years ago it was easy to dream and with determination and stamina fulfill that dream.

Our property has been an inspiration for several hopeful ranchers. The first, the original homesteader, had to be the bravest if, in retrospect to us, not the biggest gambler.

He was to buy his land for a dollar an acre - not an excessive amount for even those times. The incidence of abandoned quarter sections, even in this area, verified the small capital which was required, and the ease with which the land could be tossed aside. These acreage costs seemed just sufficient to separate the men from the boys, the stalwart from the dabblers.

Some homesteaders worked their pre-emption, then sold it- occasionally gaining a small profit - to a harder working or more fortunate neighbour. Some pursued and endured, others gave up to return to the towns, the mines, or the logging camps. John Parsons McCuddy came to our land and stayed.

It was one warm summer day in the late 1880's when McCuddy first saw the property which later was to become his home. He worked at that time, some one hundred miles to the south on the construction of the Columbia River bridge at Wenatchee, Washington.

Always alert to construction projects underway, McCuddy had come by to look over improvements being done on the freight trail that linked the newly discovered Fairview gold finds on the west slopes of the Okanagan Valley, ten miles north-west of Osoyoos, and the McKinney and Rice gold mine developments, some twenty miles north-east of Osoyoos.

At that time the lush, mosquito-ridden bottom lands of the south Okanagan were home for the Osoyoos Indians, and rangeland for the growing cattle numbers of Judge John Carmichael Haynes, then the Magistrate and Customs Officer at the border town of Osoyoos.

Since the 1860's the mild climate and tall bunch grasses on the benches of the valley had been the natural holding and wintering area for American cattle drovers moving their cattle northward to the market created from the Cariboo gold mines. Judge Haynes, using the ease of claiming cattle rather than gold or money as duty - then sending his own funds to the government - was rapidly building up a large size herd.

In the northern reaches of the 90 mile valley, Thomas Greenhow and Cornelius O'Keefe, two of these American cattle sellers, saw the land on a trip north, liked what they saw and decided to build their own herd on lands to the west, and south of the present city of Vernon. The heart of the O'Keefe Ranch still exists today as an historical tourist attraction.

Nestled between Haynes, and the Greenhow and O'Keefe ranches, Thomas Ellis had purchased in 1866 some of the earliest valley land available for pre-emption, at the foot of Okanagan Lake. Here, the hard-driving Irishman began to build up what was to become the largest Okanagan cattle herd; the basis of it, like that of Haynes, was the purchase of American stock moving north to the Cariboo. Subsequently, in 1895, following the death of Judge Haynes, he was to buy out much of the judge's lands to become the main cattle baron in the area. His 20,000 head of cattle were to run from the border to north of Penticton, a distance of over 40 miles. In 1901 his daughter Eileen married the legendary Pat Burns of the cattle buying and meat packing empire.

This was the agricultural climate McCuddy was to encounter during his first visit to the area. More important, however, than the number of roaming cattle were the activity and growth of the gold mines that dotted the lands on both sides of the border. The 1860 Rock Creek gold discovery had generated the American mines in Washington at Loomis, Conconelly, Chesaw and Palmer, and the later finds at McKinney and Fairview across the border in Canada. As the Barkerville and the Cariboo mine production diminished, these local mines came into their peak production. By 1893, Fairview became the biggest and liveliest interior town, with its competitor Spokane, across the border. Its four-storey "Big Tepee" hotel became the talk of the west. McCuddy recognized the developing potential of the area but, still involved in the Wenatchee bridge contract, he was unable to take advantage of what he saw. That is, until a badly crushed leg from a construction accident put him in the hospital. By the time the leg was functional, the bridge construction had closed down for economic reasons.

He and his wife Sarah, an American school teacher, moved north to the border town of Oroville, Washington, where they operated a store and post office while again McCuddy investigated the land across the border east of Fairview.

By this time a road linked the two mine developments at McKinney and Fairview, and the ranch site midway between could serve as a food producing and stopping place. In 1892 he took possession of the land with Sarah and newly-born son Arthur, and, in July 1893, filed for his first pre-emption of land.

McCuddy brought with him a wealth of construction experience. Although born in 1855 in Ontario, he had spent over 20 years working on railways, bridges and irrigation projects in the United States, and, to a lesser extent in Canada. He had taken on construction contracts in New Mexico and on the CPR line around Lake Superior. In addition, he had worked on the Bear River Irrigation project in Idaho and the Great Northern Bridge on the Columbia River.

Like most homesteaders, his first job on the ranch property was to build a roof over his head. This was accomplished by a solid, two-storey log house at the side of the road within easy distance of the passing freighters and stages.

All windows, doors, flooring, nails and hardware, everything but the logs, were bought from a hundred miles north at Vernon. This required that the material be shipped by boat from Vernon to Okanagan Falls at the foot of Skàha Lake. There McCuddy and an able companion loaded it onto two wagons, and hauled it south to the head of Vaseaux Lake. There they built a scow and poled the load down the lake. Returning by foot the five miles to the head of the lake, they were able to negotiate their empty wagons over the narrow trail and along Vaseau Lake to their stock-piled supplies on the scow. They then reloaded the materials onto the wagons for the remaining 18 mile trip to the ranch.

Soon the house was enlarged with another wing to 14 rooms and from 1893-1905 the McCuddy Place operated as a rest and over-night stop for the freighters, horse men and stages moving either between the gold mines, or through from Penticton to Marcus, Washington, on the Columbia River.

With the assistance of the six to a dozen hands that McCuddy employed, a bunkhouse and barns were built with a roomy, four row livery stable to provide coverage for the overnighting of the stage and freighters' horses. Vegetables and strawberries were planted for the meals at the stopping place as well as for sale at the mining camp 15 miles away. Pigs, sheep and cattle were raised as well as chickens and turkeys. A bountiful feast was ever available for the road travellers.

A 40 acre meadow was cleared of tall virgin spruce, tamarack and fir. Stumps were blasted or bored and implanted with saltpetre and burned. The rocks were picked, and finally the meadow was seeded to clovers and timothy with a natural invasion of the indigenous fescues.

During the later period of the McKinney mine's operation, high grade ore was melted down at the mine and formed into bricks. These bricks were surreptitiously shipped out by way of the McCuddy's. The gold was carried by a solitary traveller driving a light wagon. When this traveller needed to overnight at the McCuddy Place, he frequently just tossed the sack of gold under the table for the night. No one, except the McCuddy's ever knew what the bag contained.

Like most mining booms of the time, the era of growth was to be short-lived. By 1905 the production from the local mines was rapidly decreasing; the miners and prospectors were moving elsewhere or settling on the lands. With Arthur in school in Fairview, Mrs. McCuddy took over the store, the post office and the brand new telephone line linking Penticton to Greenwood.

Back at the ranch, McCuddy's herds of cattle and sheep had steadily increased in size to many hundreds of animals.

McCuddy contracted with Pat Burns, the main meat supplier of the time, to deliver the bulk of this herd to Greenwood - some 45 miles away - a railway town and the location of one of Burns' buying markets. However, when McCuddy arrived with the livestock, Burns or his representative would not pay the agreed price. Rather than be taken in by the situation, McCuddy rented a nearby ranch, bought a large supply of hay, hired a butcher, slaughtered and sold as the demands of the mines and towns dictated. Needless to say, this enterprise into direct selling of meat went on for several years, with some of the herd sold on the hoof to other small farmers and ranchers.

When life began to settle at the ranch, John McCuddy and son Arthur continued to raise their cattle and sheep with an expanded deeded acreage of over 1100 acres. Soon after Arthur left the ranch for agricultural studies at Washington State University.

Following the First World War, the premier of B.C., *honest* John Oliver, decided that the province should buy up the remaining undeveloped south valley lands, (once the Haynes and then the Ellis lands), but since 1906, owned by the Southern Okanagan Land Company. His plan was to develop these lands as a fruit growing area. Fruit had been grown near Vaseau Lake since 1896, and around Osoyoos since 1906, and orcharding was well underway just south of the border.

Planned as a re-settlement site for the returning soldiers, the Premier envisioned a Garden of Eden within the dry desert lands, and alternating swamp lands caused by the flooding, meandering Okanagan River. His grand scheme, commencing in 1921, saw the river straightened and diked with flood control dams built along its length. Wood-stave siphon irrigation flumes, and cement canals were used to service the 10 acre orchard lots.

These orchards, once planted, required several years to begin producing. In the interim while the settlers awaited their production to begin, they planted cantaloupe and tomatoes between the trees, for a short while making the newly settled area of Oliver the Cantaloupe Capital of the country.

With the new stores and services being built in the settlement, Fairview, up on the bench above became redundant. One by one its stores were closed as the new Oliver townsite took over. The post office moved to Oliver. Mrs. McCuddy closed her store and moved back to the ranch.

Arthur took advantage of the new settlement and purchased several parcels of orchard land. On one parcel, along the river, he planted an orchard and moved their small Fairview house to the land to provide living accommodation while he constructed a larger home.

John Parsons McCuddy, and wife Sarah

Stacking hay, late 1800s

Early McCuddy Stage Stopping Place, unknown rider

Forking hay for cattle

Not too long after Arthur's move, in 1928, the large log McCuddy Stopping Place burned to the ground. This was caused by an overheated cooking stove which was fired up to cook the noonday meal for the July haying crew. With the help of the crew some of the possessions were saved, but the old, tinder-dry building was lost.

By this time, with John McCuddy well into his 70s, the family decided to build only a small cottage as a replacement, and in this John was to live out his life until his death in 1937. At that time, Mrs. McCuddy moved to Oliver to be with her son Arthur, who, involved as he was with his orchard, did continue to run a few head of cattle at the ranch, but chose to winter them on his valley property.

As the original equipment was sold or purchased by collectors and buildings were dismantled, burned or had collapsed, the ranch took on the soulful appearance of many outback ranches that were visible through the countryside during the 1940's and 1950's.

In 1960 Arthur decided it was time to sell the top of the hill property to Charles Mathers of the Fraser Valley. Unfortunately, a severe heart attack ended Mather's dreams for an intensive agricultural development there. Dick Brown, a local rancher, added the 1100 acres to his valley ranch in 1964. The hayfield and acreage, henceforth, were used for late fall cattle range, as Brown continued to centre his efforts on his valley ranch. Dick finally dispensed with his ranch holdings in 1972, and the larger part of the McCuddy lands were to become ours, with our loss of 300 acres sold to an absentee American owner.

Except for a long neglected small orchard, an old log barn, a tool shed, and the remains of the cement foundation of a bay window (beyond it, the forlorn replacement cottage), some old rotting house boards with square, hand-made nails still in them, there was little left of the original McCuddy Ranch when we purchased it.

Digging wells and water-line ditches uncovered memories of the past in round-bottomed, aquamarine soda bottles and jars. Unearthed were pieces of wrought iron from the stove which the Chinese cook and Mrs. McCuddy had used some 70 years before.

Time and neglect had taken its toll. Like many other turn-of-the-century small ranches, costs had outpaced returns, and this once viable ranch had joined the ranks of the used, but no longer maintained ranchlands.

We were grateful, however, for anything salvageable on the land. In the orchard we found the apples and plums still edible, and that the crabapples made a very fine jelly. The one log barn that remained, still with its original tamarack shake roof, allowed sunlight to filter through, and wind to rattle its structure, but through cold February nights many of our cows can trace their births there and remember a snuggle in the hay, so much more comforting than the cold ice in the calving yard.

And Arthur McCuddy, very much alive, had been there to witness again a struggle with the land; a new beginning, a new dream, but this time ours.

SIX

WOMAN

Women hold up half the sky.

CHINESE PROVERB

I KEPT ORDER IN the house, in our spending, and in our heads. As old Alex, the very special Indian, used to say, "I do what I can." I could play road runner, I could pile and burn, I could fix fences, if not build them. I could dig and tend the garden, mow the lawn and do garbage runs.

Raised in the city with dogs and cats the only "livestock" I had known, I had no experience with agriculture, cows or calving problems. In fact, I rather doubt that I knew the latter existed. In a matter of fact manner, I had easily and naturally delivered a son. So I did not recognize that cows could not always do it the same.

Urban I was, and like most city persons, I considered ranching and, in fact, most agriculture was like women's work—essential but in the scheme of things not very important.

It is true, I had wanted to move to the country, to leave the traffic, the smog and evil air behind. First to the Upper Squamish Valley with basic homesteading, then to the closer to town ranch property, but I had not an inkling of the reality of ranching with such expansive undertakings.

I was perplexed to see that a tractor was as big as a tank with a strident sound nearly as loud. "You don't expect me to drive that?" but drive it I did, though never too frequently. I never recognized that the early spring smells would be that of the heavy pungent manure, and yet within the spring, the red winged blackbird did call as sweetly.

I had been a product of my time and my place, and in that time, during the early '50's, like many of the adventurous young, who were not yet jaded by television's documentary blitz, had hitch-hiked across Canada with a girlfriend, and later spent six months travelling and studying in Europe. Subsequent to that, I had spent several months sketching in a small Mexican village. I had disciplined myself through many years of university and art school; I had been employed in many demanding jobs, but I knew that I was not acquisitive nor competitive, and that I was impatient with dull and pre-ordained lifestyles. What then was I doing on a ranch?

Perhaps, being a risk taker and never being intimidated by the ambiance of life, I allowed myself to be bewitched by a challenge bigger than myself. Once planted, tenaciously I did what I could, and found I could rebuild the house.

Like many city women, I had used a paint brush and a hammer, and at least a small screwdriver; still I wasn't sure, when on our arrival I heard my voice say, as if from some other time and some other place, "I'm sure I can do most of the work on the house. I'll need help with shaking the roof, probably with the kitchen and bathroom counters, but there's much I can do myself."

I knew, having said that, that there would be no way out, for Henry so occupied with the ranch could then completely forget the house, leaving it my responsibility. And as I stood looking at the woebegone sight of devastation, I wondered if I could do it; if I alone could.

Our 12 year old son Eric had done most of the cleanup of broken glass and visible bird and pack rat nests. It was up to me next to remove the bottom horizontal interior boards and clean out the nests in these difficult areas. Then, once scrubbed out, to replace the boards and pour zonolite insulation between the studs, and face the walls with black building paper. Next, random width, rough fir boards would be cut with a hand saw, and used to face the interior, which made eight inch warm, thick walls when finished.

In conjunction with this were the window repairs: With guidance by Henry, 55 small cedar-framed window panels needed replacement. In some cases, replacement mullions had to be lathed at our local sash and door centre, then carefully fitted in place. The removal of the very tough window putty proved time consuming and exasperating. Only the nip in the early autumn morning air signaled me to persevere.

Dusty interior walls were ripped out and relocated, and the stairwell wall pulled out to give more openness and light.

It was disappointing to see that the relatively solid three inch fir flooring would have to be covered. The house crawl space was too low to work and insulate within. Consequently, insulation would need to be applied to the floors themselves. Styrofoam insulation sheets were used, with 3/8 inch plywood sheets placed on top, then carpeted.

During that first autumn of rebuilding, I was fortunate to have the help of a concerned clerk at our local lumberyard. In many instances it was he who cut my plywood to fit awkward corners. This job of cutting—with the use of an often too dull hand saw—I had failed miserably at doing.

At least once a week I drove the Land Rover into the lumber yard to pick up my carefully measured materials. There with the clerk's help the pre-cut plywood, insulation and lumber were tied to the roof racks, and I drove up the hill to work until time for the afternoon run back down to the school to pick up our son.

With Henry's constant observations, in this pushy constant way, six weeks after my beginning, we could move from our cramped 20 foot borrowed trailer into a house, still without plumbing, but with sink and at least cold running water, if not yet hot, and with counters and shelves, an Ashley wood heater and solid windows and doors. The Dutch back door I constructed from 3/8" plywood, faced on either side with 7/8" cedar boards. It was my creative pride and joy.

Our personal belongings and the small amount of furniture that we had brought with us had been stored in the one solid building, a relatively new BC Tel storage aluminum-faced garage. Gradually these items could be moved from there into the house.

With the use of cedar shakes that were cut on our coast property, the roof was re-done, in this case (as I had hoped) by the three of us. By Christmas our little house, warm with the glow of the wood interior, was equipped with a propane built-in range and refrigerator, both of which helped to handle the warm festivities of our first ranch Christmas.

Through the winter and later into spring, we were not without our previous tenants as callers. One determined, dispossessed pack rat scurried each evening to the glass of our French door to stare in longingly at the warmth of an interior which he had once called home. Finally, discouraged by the lack of entry, he sought and found alternate shelter in our small, wooden tool shed. There, the following spring, we found his carefully aligned selection of nuts and bolts, nails, screws and fencing staples.

That spring the house wrens beat themselves aggressively against the glass of the back hall window, convinced, I suppose, that if they beat long enough somehow they would locate the hole that used to be there. The swallows in comparison seemed content to hang their nests against an outside wall and raise their two groups of offspring as they had before.

As soon as the snows receded in March, I returned to the paneling and finished the downstairs and difficult peaked bedrooms, and alcove of the small upstairs. By opening up the rooms to the full pitch of the roof, the rooms seemed larger and more usable. Here, however, the thick fir boards had to be nailed horizontally to the overhead studs. This job seemed a nearly impossible task for one person, especially when that person was me.

Determined, with my hands slivered from the rough, dried to the hardness of oak, winter-stored boards, I stretched out on my back, pressed each board with my knees against the lower extent of the sloping studs, so that my hammering could get that first nail firmly into the stud. Once the board was held by the first nail, the finished nailing was relatively simple—except, that is, for the iron-like hardness of the boards. I cursed as I bent nail after nail.

More evident in the upstairs was the house's lack of square. This required measurements for each board to be taken separately, then cut and hauled up the stairs to be nailed. Like climbing a steep mountain, it tested my patience as I mismeasured time after time, thus requiring that I duplicate my efforts.

The oblique angles for the gable ends of the rooms troubled me most. In desperation I discussed this with Henry's father, who while on a visit, had ably built the kitchen and bathroom counter framing. Within days, in the mail came a sliding bevel. This dandy device went a long way to simplify my angle measurements.

To make the best use of all the available space, I built book shelving and storage within the recess of the eaves. We left unchanged the random-width fir flooring of the bedrooms, except for staining and waxing.

We had brought with us from the coast a good supply of finished cedar boards. These mill ends, then considered nearly castaways at $30. a thousand board feet, became our cupboard doors, shelving, counter tops and even bookcases. To this point we had spent under $400 in materials, the largest outlay being for insulation and carpeting. With the house wiring, the bathroom, washer and dryer yet to be added, we would be well within our $1500. budget for re-building. Rough fir lumber for exterior board and batten siding had already been obtained from the sawmill which operated for a short while on our property. On that work I would proceed the following spring.

At that time, living in the cottage was considered as a short term occupancy. Within a couple of years we optimistically planned that Henry would design and have built for us a new home.

Some foreseeing of the future, or my caring nature, drove me to do the best and most crafted job possible within the material limitations. This frequently meant pulling down an area and redoing it. Henry's critical eye drove me if I relaxed for a moment on my own standards. Problems were solved as I came to them. Cedar laths, sanded and waxed, were useful for facing irregular-sized closet doors, and the functional window casing, and beading proved indispensable for camouflaging the rough edges of my carpentry. Though the house would never be a show place, it possessed a unique charm which trained carpenters, and a large budget would never have obtained.

Looking back, I appreciate that I could do this, thus leaving Henry free to do work which he could do, but I could not. I was grateful too that Henry encouraged me to concentrate on the house and yard, not expecting, but appreciating any help which I was able to give him for feeding the herd and assisting with the inevitable emergencies which did appear.

I was quick to realize that I could not compete, nor even adequately assist, in this macho world. Yet, I stood in awe at many farm-raised wives as they disced and raked, threw bales and forked manure. In thought I defended my frailty; perhaps, if we had started all this 15 years ago. And she does have a daughter, who does much of the inside work, and they do have all this easy, flat land. In some instances, I had to admit that they had none of these, but worked equally hard, equally long beside their husbands.

Eventually, in questioning these physically active women, I found often the same discontent that exists in any working wife, who must not only do a demanding day's work, but miraculously carry the complete responsibility for food preparation and maintaining the home. Often, it was the rancher's dream, not her own, which kept them a near slave of the land.

Sometimes, more often amongst the young, the ranch wife handled disorder in the house easily and preferred to work outside, treating the house as just a shelter for a quick meal and a sleep. With a similarly-minded husband, even within the confusion, a reasonable state of harmony existed.

Henry's love of fine food, served within as gracious as possible environment, made it easier for me to follow my priorities. Nearly 20 years of youthful wandering, with as many addresses to call home made rooting more significant to me. A home, however modest, was suddenly enduring and singularly important. Perhaps more so for me, because the ranch property seemed so boundless, and so racked with demands for attention, that I needed an order, a controlled design within our home. I was not inclined to abandon this easily for field or cattle chores beyond my time or capabilities.

Contending with the permeating confusion and problem oriented ranching life did not come easily to me. By the grace of God, or by my capability and volition, I had usually been able to be on top of my life, thus I found I disliked an environment which made me feel inadequate. How could I directly solve the problem of a bloating bull or a leg-back, calving heifer. I could call for help, point out the problem and assist, but never directly solve the problem.

It was to take many years for me fully to recognize that there were certain weeks: range turn out, the frenzy of haying, when I could not ask for nor expect, anything from Henry. It was to take me many winters to cease dwelling on the fact that only a pump and an electric heater stood between our having a water system and losing it. It was to take me many years to recognize that I need not feel guilt because I could not rush out to help every bawling calf locate its mother. This dependency on human time and human error, rather than urban technical systems, confronted me; this newly discovered inadequacy and helplessness bewildered me.

Where had the solitude I once depended upon fled? The ubiquitous problems infiltrated all thought and action. I could not insulate myself against them.

Unlike Henry, I was not by nature a doer. For him it was difficult to sit still. His years at a drafting board had caused him to find distaste in more than one or two hours of desk work. He was happiest when actively involved in a project—the bigger, the better.

Through my early, but I admit satisfying years, with building, beautifying, piling and burning, I could keep far from the heart of much of the stress. As paper work mounted, I became a sifter, a filer, an expediter, the centre for all ranch communication. Always, I saw my equal share of the awesome responsibility for keeping the ranch going.

Constant change and newness in itself can be stressful—and our ranch was always changing. No sooner had we devised and built an operation with chutes, holding pens, feeding techniques, than our herd numbers increased and our hayfields became larger—facts which required a complete revision of everything that had come before.

Perhaps, easier it could have been, if our move had been to an established ranch, which changed only with the seasons, not drastically with each year as ours did. Or perhaps, had I been aware of less, saw only the seasons, separate from the ranch activity, but I could not. Winter had cast aside its long association with skiing and firesides, spring was no longer primarily a melody of sights and sounds, summer lost its previous meaning of easeful swims and sun, and autumn seemed unrecognized in all its misty splendor.

My thoughts were never clear on why we were struggling here; they were never clear on why I was doing this. Sometimes, I could voice some of the reasons, but was subjected always to the pressures of change, to the immediacy of how. Without the thought-time to question fully why, I pushed on.

I knew why we had come here, but the reason escaped me why the ranch project had become our life; why I had put aside my life goals, had abandoned all the previous ways of seeing, being and loving. Had we, so soon, gone too far to turn back to how I had envisioned it: a casual life, a peaceful place in the country. Certainly we had fewer opportunities for walks and talks, for exploring horse rides or social occasions; instead, the years of effort brought only work, and more work.

I questioned what there is within the human spirit which causes us to catch the spark of an endeavor, and drives us to make it flame? What is behind those oil men like Jack Gallagher, human dynamos like Ian Sinclair that caused them to push forward? Is it solely the need for power and control, or does the growing act itself, the process, become an energy with its own momentum? Or is it that any form of lifestyle after the 'grand slam' is a form of death? For us, after the fervour of ranching, an ordinary job would bear a very tarnished comparison; idle, carefree days, supported by bank deposit interests, would be close to parasitic.

I could argue, of course, that it was the obligation to our son which motivated us. Within ourselves we wished to build an economically viable ranch, but not only for ourselves. We wanted it to progress again through another generation.

Through his growing years, Eric's commitment to the ranch had cost him much in terms of companions, play and athletic activities. He, too, was caught up in the flame of accomplishment and put aside his childhood dreams of archeology and astronomy for college studies in agricultural management. Could we abandon him without a choice of a future here to the work of managing another's farm or business?

I could argue, for us, that our obligation was to produce food: good wholesome beef to help supply the nation's need. But that argument began to diminish when the passing years forced us to recognize the consumers' growing resistance to animal products. This was initiated by the introduction of margarine and vegetable oils. That early research linked the consumption of eggs, milk, butter and meat (fat) with heart and circulatory problems. Though later research proved that the earlier findings were inconclusive, the damage had been done and continues to be perpetuated by the advertising and media dollar. In turn, many of the confused consumers believe that their health will be safer with solvent-derived margarines and vegetable oils and chemically simulated animal products.

I could argue an obligation to the land itself, to the ranch itself, and to our growing community here - an obligation to continue the hub of activity which the ranch generated. But that again would be only part of it.

As nature does not show itself as a single face, our reason for pushing on could not either.

SEVEN

THE BLUETONGUE CONCERN

Conflicts originate between men when they do not say what they mean, and do not do what they say.

MOUSTAKAS

THROUGH THE TOTALLY engrossing development years of 1975-76, we were aware of, but tried to ignore or belittle, the rumours of a cattle Bluetongue outbreak in the southern interior of the province. Stories circulating were that, following roundup in the fall of 1976, cattle would all be tested and the suspected cattle would be slaughtered.

Off the beaten track as we were and with only radio telephone linking us to the outside world, we had heard nothing direct from the Health of Animals Branch, therefore preferred to deal with our immediate concerns, rather than worry about later claims upon our time. However, by late summer the truth of the government's demands for our attention and cooperation again had to be faced.

Like many other cattlemen, we resisted having our cattle move through the chutes unnecessarily, insisting that our cooperation with their testing program must coincide with our fall routine pregnancy testing chute work.

By late October we could no longer delay this, or the inevitable arrival of the technicians to draw blood for the Bluetongue testing. With no apparent illness in our herd, we were convinced that the whole exercise was one of busy work for the technicians, an effort which would lead to a clean bill of health for our herd.

A week later, we were to learn otherwise. Eleven of our best animals were suspected positive. The very top of the herd, including our herdsire 'Pollar' and Eric's four year old pet 'Inky', an Angus/Holstein, were marked for slaughter.

'Inky' had joined Eric as a young calf; her dam was a top producing dairy Holstein and her sire a son of the great Canadian Angus bull 'Canadian Colossal' who had given so much growth and size to the breed. But outside the merits of 'Inky's' breeding, 'Inky' was more than special to us. Not only did our son move her about by mounting her back, but her quietness was such that she could be vaccinated, even artificially inseminated by simply tying her to a post or tree. As a mama she did more than her share in raising not only her own super calf, but, if necessary, an additional orphan, loving and caring equally for both of them. Cows like 'Inky' have their disadvantages on range operations, yet every ranch is blessed by having at least one 'Inky'.

As there was, at that time, no other polled Charolais breeders in B.C., with the nearest breeder in Alberta or Oregon, the replacement of our fine, upstanding 'Pollar' would be time consuming and costly.

At the time of the outbreak we were not alone in our complete ignorance of the disease. Reading all the inadequate material we could find on Bluetongue, we learned that this virus disease rarely caused any noticeable symptoms in cattle. Carried by the bite of flying gnats of the genus Culicoides, it infected deer more severely, and certain of the dozen strains of the disease could kill sheep. It was not contagious.

Testing for the disease was done by a microscopic examination of a blood sample. This was done in an Ottawa laboratory by using a complement fixation test. Suspected animals, or reactors, were those carrying a high antibody count to the disease. All this, with no noticeable disease symptoms, much like Tuberculosis immunity in humans.

So why, we argued, were these cattle being sacrificed because of an immunity they now possessed to the disease? Logically, this immunity made them more valuable to a herd.

If we received any explanation from the Health of Animals vets, it was that there was no inexpensive test to distinguish between the active disease and antibody reaction, therefore it was expedient to identify and slaughter all antibody reactors. This may have satisfied the government officials, but it was too lame an excuse for us to accept.

It was natural for all ranchers, to question why this clean up of an innocuous disease was taking place. They knew, as we knew, that the disease was a common, uncontrollable one in the United States, particularly Washington State which bordered our most southerly ranches. If the disease existed there, what was to stop a reoccurrence through deer, flies and cattle which did not recognize international boundaries? Would we be expected to live with this testing, and herd desecration every few years? No one we asked appeared to have adequate answers to allay our fears.

We did learn that Canada, until this outbreak, had declared for itself a Bluetongue free status. This disease-free status gave her the natural ease, unlike that held by the United States, to export dairy semen to the sheep producing countries. In short, our cattle appeared to be sacrificial lambs to the export market. This, too, did not sit well with us, and we resisted any efforts by the Health of Animals Branch to bring appraisers to evaluate our reactors for compensation.

During this period, news releases on the outbreak were scanty. But when released indicated that the ranchers had been most cooperative in the herd testing program, that the extent of the outbreak was known and that complete eradication was within sight. The media did not pick up on any individual rancher's point of view and took as fact the one-sided release from the Health of Animals Branch.

We shuddered at the ranchers' apathy, but understood it. During the previous three years, cattle prices had been at an unprecedented low; cattle surpluses existed in all beef producing countries of the world. It was easy for local ranchers to tell themselves, "We have to cut back the size of our herd anyhow, why not allow the government to do it? They'll pay more for them than the auction will."

Locally, the cattlemen's association worked mainly toward obtaining some compensation to support lower prices at the area's livestock auction yard. These lower prices were caused by buyers' fear of handling or dealing with the bureaucracy involved with cattle bought through that yard. The busy association had to leave the battle of the affected herds to the individual rancher and the Health of Animals Branch.

But there were pockets of rancher resistance amongst those affluent enough to see beyond the practical aspects, and those, like us, seeking real compensation for our sacrifice and identification of the suspected strains of the disease in our area.

Confused as I was by the ranchers' lack of open protest, I could no longer contain myself. Our local newspaper gave bold heading status to a letter I wrote to the editor about the previous inaccurate news releases regarding the ranchers' satisfaction with the program. I gave the facts of the disease, facts which had not appeared in print, told of many ranchers' dissatisfaction, not only with the compensation, but with the program itself. I wrote our CBC agricultural noon hour show, wrote an article for the Cattlemen's magazine; which they preferred to re-write in the form of a letter to the editor, treating it as 'after the fact' information. The Canadian

Cattlemen's Association did not voice an early opinion other than print a bland, explanatory much-delayed news report by Chris Mills, the Secretary to the Canadian Cattlemen's Association. They did not respond to the proposition of which we were very much aware; that healthy cattle were being slaughtered, not like contagious and infectious cattle with Brucellosis, or in earlier days Anthrax or Hoof and Mouth Disease, but healthy cattle. This was an impetuous program costing the taxpayer up to $3 million dollars for a disease that the government had not proved to exist.

I saw my efforts to inform and voice alarm die with a whimper. To my knowledge, no one openly joined my protestations; no one had the time, energy or desire to rock the boat of bureaucratic decisions.

We were still a newcomer to the livestock industry; were we being naive to expect cohesive support for the unfortunate ranchers with reactors, from the more politically aware established ranchers who found themselves free of reactors. After all, it was not as if we were not being compensated. In fact, the Health of Animals Branch assured us that, as soon as we allowed breed specialists to enter and appraise our 11 reactors, reimbursement would be swift.

With winter soon upon us, we reluctantly consented to allow the appraisers access.

Vivid in my memory still remains that late November day. Nothing was cheery in the faces of the officials exiting from the cars, which had arrived as a cavalcade of strength. In their dark rain coats, and in the high polish of their black shoes, the aura they cast was not unlike a Gestapo of uniformed power. Having reached the gate of entry, there was no way that they were going to cast anything but a voice of seriousness over the whole business. Huddling and whispering in a group, they attempted to shroud, at the same time bolster, the tottering, stooped elder in their midst. By this act they seemed to be trying to camouflage their weak flank in this elder, the Angus breed's representative. The only sunlight came from the bright, friendly face of the Charolais breed representative who stood separated from the tight bureaucratic group. He'd arrived minutes earlier in his own vehicle.

The slow trip through the fields was set by the tottering Angus representative's hesitant steps; the group stopped, pondered, and whispered in the sight of each reactor animal. As Henry pointed out the merits or pedigree of each animal, they as quickly pointed out its shortcomings. For me, it seemed a charade, an exercise in futility, as it seemed apparent that in their solidarity, they already had come to dollar decisions.

Later, inside over coffee, this became an actuality; their inflexible compensation decisions would not truly take in replacement value. They stood steadfast on the assumption that we could go to any local auction, any day and pick up replacements. They would not acknowledge, as a financial consideration, our isolation from other polled Charolais breeders, that we would have to travel up to one 1,000 miles to find equivalent replacements.

The added rub was that the Charolais breed representative's appraisals were deemed too high, and that no attempt was made to negotiate between their low and his high appraisals. Instead, they tossed out his appraisals—the appraisals of *their* chosen representative—and informed us that they would bring in another, more objective Charolais breeder for re-appraisal.

Indeed, this was another costly exercise in futility, for when the following week the second Charolais representative did arrive, along with the supportive officials, his appraisal was to be as we expected, identical to the group's. This was their final offer.

Angry, standing firm against any attempts by the Health of Animals Branch to remove our reactors, we delayed their attempts until threatened by RCMP back-up support. By then Henry had met with the newest member of the government team, a department vet, brought in from the prairies to smooth away the last contentious cases.

In his discussions with him, Henry had become somewhat pacified by the appeal route still available to those dissatisfied ranchers such as ourselves. With an alternate route still open to have some justice done, we consented to have the cattle removed. The cattle would be trucked to Vancouver and slaughtered, and the meat distributed through the normal channels of the packing house. As I have pointed out, this meat was fit for human consumption, scarcely diseased in the true meaning of the word, yet 1500 head were to be called diseased and taken.

No attempt was made to hold and contain a group of these reactors so that the disease could be further studied. At least this would have been some reassurance to the ranchers that the Health of Animals Branch was attempting to learn more about our disease; that, in the future they would have a greater understanding, and could avoid the rashness of unsubstantiated evidence if another outbreak should occur. As the infecting gnat, for some unknown reason, seems checked north of the 51st parallel, this group could have been contained there, isolated from the southern outbreak. Critical questions might have been answered by a test group. Instead, the government appears to have forgotten that Bluetongue ever was, until it appears again.

The appeal route, under the Registrar of Appeals, Federal Court of Canada, was not simple. It was time consuming, but for us, necessary. We prepared our own brief during the winter of 1977 and sent it to the Registrar in Ottawa.

That fall the first of the appeals was held in Penticton.

I was reassured by the intelligence and interest of the Ottawa judge presiding. Though not knowledgeable in the cattle industry, he made every effort to understand, not only the industry, but to understand the concepts of the established herds of many B.C. ranchers. These herds in general were closed herds with extensive culling, and performance improvement built into the operation. They were not in and out operations often found in the prairies and Ontario, not disparate cattle purchase operations, but consistent developed herds, in some instances over a 100 years in the making. These reactors, therefore, could not easily be replaced by shopping at the local auction yard with primarily cull cows. These were breeding stock, familiar with their ranges, not a haphazardly collected, untested, unrelated herd, and, therefore, were more valuable. To our delight, the case for higher compensation was won by the appellant; they would receive a significant, higher settlement.

The ranchers for the second hearing was held some months later. In this instance, a different judge and different defense lawyers were to hear the mainly minor (with one exception) claims of five appellants.

The rancher of the first hearing had retained a lawyer to argue her case. The second group of appellants, encouraged by the results of the first hearing, influenced by the small amount of their claims should they win an award, and guided by the suggestions of the much respected judge of the first hearing, did not retain lawyers. Right from the beginning of the hearing, this was to be used against them.

The fidgeting, restless judge drifted easily into lapses of memory and, having no knowledge of cattle, or seemingly any desire to learn, soon grew impatient and testy. The government lawyer, better prepared for this hearing and more astute with procedure, took every advantage of a befuddled claimants' arguments to belittle or reject their points of view. Criticizing their inadequate preparation, their lack of supporting evidence or witnesses to defend their case, the judge quickly tossed the five cases out of court, one after the other.

On this day of hearings, even unique, but personally crushing injustices were denied by the judge.

One elderly long-established rancher who appeared in court that day, had his claim for more compensation based on the injustice which he had received, because of his early cooperation with the testing program. During the very early months of the outbreak, he had willingly allowed his herd to be tested, while other ranchers avoided it. At that time, well into his spring calving, 90 reactor cows were identified. Within weeks, these cows, and in most instances, newborn calves, were trucked away and compensation was paid on a 'cow with calf at foot' basis.

Many of these newborn calves were likely free of the disease (as our herd testing later proved many to be), yet they were shipped as baby calves, with insignificant compensation paid for them, and for the rancher's winter work and expense in caring for their mama cows. When autumn arrived, and the fall calf sales, the rancher did not have the usual income from these calves, calves that would have been raised cheaply on the ranges.

As the rancher futilely pointed out, "I am not in the bred cow or cow/calf pair selling business; that early testing cost me about $20,000."

He was also to lose approximately 300 pounds per calf in Income Assurance compensation which was then still in effect.

In contrast, the less cooperative ranchers who evaded the testing program until fall, received specific compensation for their reactor calves. In many instances, they found calves from reactor cows free of antibodies, thus did not lose them at all.

When the judge's decisions were made, and were received by the appellants some months later, there were no awards. The ranchers had lost their cases. To their relief, however, all court costs were absorbed by the federal government.

There were now four cases left to be heard. The first was the appeal of a much-respected rancher with an excellent herd and a strong community image. The second was the appeal of one of the area's larger ranches. It was within this herd that the first Bluetongue reactor had been identified, and it was he, who denied Health of Animals Branch officials access to his herd in the spring of 1976 and had, against their instructions, turned his herd out to his vast ranges. Later, the Health of Animals Branch was to hire riders to bring his herd of many hundreds in for testing. This was, as ranchers knew, a time consuming effort. This rancher's earlier contentious behavior, as well as ourselves, and the other ranchers, did not sit comfortably with the government officials. We were aware of our confronting reputations and knew that, if we had any chance to win our cases, our tactics would need to be flawless.

We questioned whether we knew enough about legal procedure to handle the court case ourselves. Could we learn fast enough? We argued the merits of hiring a lawyer, to find the best, to share his efforts and costs amongst the four of us. Time and separating distances between the ranchers made us hesitate to do this.

Our contention and case, was based on principle: The disease did not exist. No virus was isolated; no symptomatic animals were discovered. Had the Health of Animals Branch made an error? Did they have the right to slaughter reactors? The antibodies had been identified in local domestic sheep, but contrary to expectations, these sheep had not been sick nor did they die. Local deer had told the same story.

Though our financial claims were not large, our case had to be based on the most valid points. It would require the most costly witnesses to counter-balance the witnesses we knew that the government would—without a thought to cost—bring in from afar. It was no longer an appeal. It was a fully fledged battle. We were now the ones on trial.

We had witnessed the all-encompassing control of the Health of Animals Branch and the Contagious Diseases Act, and one of the last four appellants, independently, had had blood drawn from his previously identified reactors, and the samples sent to a private U.S. laboratory for Bluetongue identification. Tests there were routinely carried out by the lab, but when it learned of the Canadian Government's involvement, it refused to release its findings. Doors were being rapidly closed on us to prevent our reaching the truth about the disease, or the reliability of the testing procedure.

In February 1978, 15 months after our reactors had been appraised, we received notice from the courts that one month hence, our case was to be heard. This was to be on March 8, right in the midst of the calving season. As we expected, the bureaucrats were playing their cards well.

As we pondered our options, we received a letter from the Justice Department. This read in part that they would seek total costs should our appeal be lost.

This had to be the final blow. With everything now to lose, could we proceed? The Crown had a 'no cost too great' determination to win. How could it lose? How could we win?

We were too involved, as were the other three appellants, in separating distances and calving to find the time or energy to proceed. A week before the case was to be heard, we did withdraw from the contest, as I'm sure the Health of Animals Branch expected us to do. The remaining three ranchers did the same.

Later that year we were to drive to Oregon to purchase an unproven, bull calf replacement for the proven "Pollar' bull we had lost. The settlement received was adequate only for a calf. We absorbed these travel costs, but we could not afford to travel immediately to replace the other reactors.

EIGHT

RANCHERS ARE BORN, NOT MADE

The dream of man's heart is that life may complete itself in a significant pattern. Vocation gives us such a pattern.

SAUL BELLOW

THE TRUTH IS that ranchers are born, not made; that their wives are also born, not made. Certainly, any number of men and women can learn the skills of caring for cattle, as well as the required farming skills which can be learned in a classroom, but this born rancher in today's society is an anomaly—a 20th century man guided by 19th century principles and ways of acting.

Like an artist, he must have a capacity for pain, for enduring, for hanging on to his dream of a better herd, a better ranch, holding on when all evidence says he should quit. He gambles through the depressed prices of an over-supply cycle, optimistic that the turnaround in prices will be significant enough to off-set his losses. It rarely is.

Like a woman in the labour pains of childbirth, he soon forgets distress and like her, soon pregnant again and optimistic about her forthcoming child, he forgets and plows his efforts back in again.

His experience is with the unpredictable changing seasons, and within these vicissitudes, he easily comes to expect a similar condition of high and low seasons in cattle prices.

The insidious nature of the cattle business, the ups and downs of cattle values in one sense protect him from a complete recognition of his own worth, and equally so protects his banker from precise analysis of this worth. Figures, if not lying, can at least mislead, particularly if given an impressive sales pitch by the rancher. The supply and demand marketplace with its fluctuating cattle prices can indicate in one month a healthy debt/asset ratio, but the next month with cattle values cut nearly in half, this ratio can be severely out of balance. Thus the banker, traditionally, has loaned on the individual rancher's past performance, on his ability to 'pull it off.'

The disparate traits of gambling and enduring are not solely exclusive to ranchers. Any enterprising businessman or entrepreneur must gamble and endure equally; however, the rules of business must be more soundly adhered to in the urban business scene. Beginning losses can be tolerated, but the small businessman cannot accept severe cyclical losses, for his profit margin must be excessive to tolerate these periodic losses, and he would lose business. By producing an essential product with no predetermined prices, here ranchers separate themselves from true business enterprises for, although ranching is a business, it is not a true business. It is doing what one wants to do, a vocation for which a certain kind of person will sacrifice much.

If one considers a rancher's investment in land, equipment and cattle, it has never—in recent history—shown a position of excess profits. In fact, during the past ten years, even with the cattle Income Assurance amounts, ranchers would have experienced only a few years with a small return on his investment, and this would soon be devoured by the losses of other years. Beginning ranchers may have covered their operating costs, but cannot have covered both principle and interest on debt. How then have they been able to continue?

In some parts of the country, land inflation has escalated higher than in other areas. Ranchers with land close to urban areas or within recreational areas have been able to borrow more easily against this 'retirement fund.' Other ranchers have had to work part of the year off the ranch, neglecting cattle and ranch, or to survive have had to send their wives back into the work force.

Every rancher today knows that if a sum of money equivalent to his ranch investment were earning interest in the bank, it would give him an income far above that of the average worker. Sitting in an easy chair, his feet at rest on a stool, he could spend the next 20 years on recreation, hobbies and travel.

Then why do ranchers not quit? Why do they not take the easy way out from an obsolete way of living? Why do they rise to face each new emergency, grunting, groaning and often aching from years of neglect to knee-caps or backs, but still rising to the occasion and going on? Why? Because they are tough; and because they are doers, and because they are men of action, not indecisive, intellectual game-players, and never by choice quitters.

Why do they continue to eye the correct pattern and colour of the Hereford bull that they are about to purchase? Why does that uniform herd matter to them? Why do they want to gaze out at that pastoral scene of balance and uniformity in a particular breed, when specialists have told them that cross breeding and hybrid vigor are the routes to greater poundage?

One has to look back at the mythology of ranching to begin to understand the contradiction and the ambivalence of the rancher.

Within the visionary mind of man, the fantasy mind of man, the spread, the herd, the cow/horse have always represented a freedom, an anarchical posture of self-sufficiency, a life-style separated from the greed of business and the intervention of government. The ranch has represented a world unto itself. In many instances, larger ranches have become towns unto themselves, complete with store, service centre and school. This initiative to do one's own thing is still an essential trait in man's deepest soul. For the rancher, part of that total entity is to be surrounded by the aesthetic form which suits his nature. For some, it is the expensive spirited quarter horse; for others, the neat white fence around a manicured lawn; for others, it is the gracious attractive wife. It is not the profit, but the chance at a controllable lifestyle they choose, the empire of a man still free. Because of their determined inflexible natures, they have survived. Perhaps they have survived for only these reasons.

Those raised on the ranch separate themselves by having learned an uncanny way of coping—one that, I doubt, at this late date we can ever learn. They no longer see their day to day experiences as problem; instead the uniquely unexpected events are the only problems. The problems of calving are not problems, but the way of calving. A problem there is seen only as a problem if it has some new way of showing itself. For instance, if a heifer at calving loses her calf, this is just the normal odds. But if the heifer has the calf half-delivered, then because of fright swings herself around and in so doing bangs the calf's head against a solid wall, thus killing the calf, that is a problem—one to ponder over, perhaps even to share.

I said, too, that ranchers' wives are born, not made, for indeed their endurance and gambling natures must be as great. Their love of the land and nature must be an intrinsic part of their priorities. More important, they must have the ability to handle chaos and heartbreak, to remain patient through long years without holidays or weekends off, to have a self-contentment which requires little social support. A woman has not only the stress of the ranch, but her husband's stress to handle as well.

Most women not born to the disruptions, the confusions, and problem-oriented life cannot adjust to a life different from the quiet, routine life that she has known. For perhaps only this reason the ranch could soon be back on the market.

When I began to meet long-time experienced ranchers, I was first struck by their slowness of movement and speech. In time, I began to understand why they were like that.

Like comparing dermatologists and obstetricians, there is a critical feature of livestock producers, not as frequently encountered in grain, or ground crop or fruit producers. As with these producers, the ranchers' seasonal stress is there, but there is as well the more troublesome stress of frequent crises in livestock.

The more controls that the rancher wishes to place on his animals, the more problems he encounters. Animals can never be commanded to stay put as an apple tree can. They can never be ignored, or closed down as an orchard or crop can. They forever represent the weakest link in the fence, the potential for disaster.

Unlike the traditional farmer who has his peak planting and harvesting periods, and his 12 to 14 hour days during these times, a rancher has no peaks even though he sometimes feels that calving and haying represent equivalent peaks in his work load year. In reality, these are just high activity points of a steady, yearly, productive cycle. If he takes his work seriously, if he is still relatively a beginner, his work load can be a dawn to dusk assignment, a 12 month occupation.

The family-sized ranch (100 to 150 cows) with perhaps only one full-time hired hand or perhaps none, can have 10 to 12 miles of fence to maintain—a fence which, at least in B.C. , in the main, runs through forested land and is thus vulnerable to cattle and bear breakage, to indignant motor bike clippers, to windfall tree damage, and also to the usual five to ten year rot-out of posts.

The rancher has cattle to check on mountainous ranges, salt to deposit, cattle to move from persistently overgrazed areas to lush but more inaccessible areas. Crushing his spirit, he has the chronic problems of gates left open by tourists and hunters, which often leads to days of riding to collect and return cattle to their proper areas.

In dry summers, he must investigate his water holes and, if necessary, find and develop additional water sources to cope with the shortage. Within these demands, as often as once every five years he must re-seed his alfalfa fields and yearly cut, bale and stack his several hay cuts. Low precipitation areas have the additional demands of irrigation, the usual twice a day burden of sprinkler line moves.

In late fall, he has the round-up: the sorting, weaning, selling of calves and cull cows. This is followed by days of riding in near winter temperatures to try to locate strays without solid evidence that they are alive or even there.

Lastly, he has a fleet of cultivating and haying equipment to repair and maintain; in numbers if not in size as great as those of a grain farmer.

In contrast to the orchardist, the grain or crop farmer, he does not experience the relative ease of winter, for winters, in conjunction with his projecting, accounting and, equipment repairs, he has the feeding and health care of his livestock. By February or March he is up in the night, sometimes through the night to calve out and oversee his newborns.

Within this routine, the rancher soon recognizes that everywhere he looks is something to be done, and all that is to be done is dependent on the amount of energy he has. As the years pass, suffering as he does from a slow but pernicious burn-out, he learns to move slowly, to conserve, not waste, his energy. Without a resurgence of energy from a rarely taken holiday or a complete break from the ranch, turtlelike, the rancher moves through the day, decreasing his objectives and putting on his blinkers.

Later some of these ranchers surely push on out of habit, as marriages often push on out of habit without the energy or wit to know how to undo them. Their endeavour becomes an effort outside of sense, perpetuated on its losses with alternate enterprises not pursued. The rancher stays on, so deeply implanted that he, in all honesty must ask, "What else do I know how to do?"

Ranchers, too, have been known for their privateness, their suspiciousness of those not born to the land, to its traditional roots of behaviour. Our first meeting with a rancher outside our immediate area was to confirm this.

Parking the Land Rover, we cautiously inched our way through the mud of the rancher's long driveway. Yet not conditioned or prepared for the spring mud of most ranches, we had worn, not boots, but shoes, that made our approach look as ridiculous as that of uninitiated city visitors. If it had not been Sunday, we surely would have been taken for sales people.

We picked our way slowly, observing and delighting in the newborn calves at play, their protective mothers sunning, scratching themselves against the side of an animal shelter.

"Look at that magnificent barn! Oh, if only we could have the time to build barns like that."

"Or have the luck to have had those barns on our place," was Henry's reply. We bemoaned the high cost of our loss, the fortune of this rancher, a third generation resident, who had a father with the foresight to build these crafted buildings, constructed as if they were to last several generations.

"But I see who was the boss," I added as we caught sight, not of the same skillfully built home, but that of a very ordinary, recently built ranch house.

"Maybe it burned down," was Henry's reasoning, as we focused our attention on the tall, slightly stooped man coming from the house and through the gate of the picket fence which surrounded it.

As he neared us, Henry started to put out his hand, then as quickly withdrew it as he caught the man's expression, that of a man coming to see us, not to greet us.

"Ted DuMont, I presume. I'm Henry Mann, my wife, Elizabeth, from over the hill."

"Oh, is that so," the man mumbled cautiously.

"Yes. We were wondering about joining the Farmers' Institute. We need to buy a lot of barbed wire and cattle supplies, and though we're not really from this area, we heard we might be accepted as members."

"Well," he drawled sardonically, "there are quite a few city people coming here and buying up 160 parcels, lasting a year or two, then going back to where they came from."

He did not add, "Like you," but it was very clear to us what he was thinking.

Henry was beginning to grow impatient with the indifference of the man towards farmer support.

"Think you're going to be able to make a living there?"

"We're going to try," was my quick reply. "We do have 800 acres and nearly 1000 more for grazing".

"Oh well, that's a bit different," he replied, his interest somewhat perked up. "But I'd sure hate to have to drive that road."

"After city traffic, it's a pleasant change." Had he heard this all before, I wondered.

Henry was ready to go and I knew it. It had grown cold in the damp late afternoon air, and I shifted impatiently on suddenly very cold feet. I was ready and willing to join him.

Hesitating only briefly, I heard Henry's determined voice, "Well, there is no reason why I can't drive to Vancouver myself to pick up the wire."

And as we turned to go, the rancher said, "Wait here a minute." We obeyed, shivering and lingering as he entered the house, returning to say, "Will you come in?"

Chilled and stunned, we meekly followed him into the warmth of the living room fire.

Having made it across that magical threshold, we were treated like very special guests, and were warmly welcomed by the rancher's charming wife. Within minutes freshly baked pie and coffee were served. As we warmed ourselves, there followed a long discussion of our ranching objectives during which even the words ''Charolais cattle' were mentioned without the expected cooling of the mood; a sharing of their ranching experiences continued, then our name was duly registered as an institute member and attention given to our order for supplies.

From that day onward, accelerated by their return visit to us, this ranch family was to have a very special meaning in our ranching lives. Through them, for the first time, I saw with some degree of awe the nearly unique human traditions still symbolized by a rancher.

Apparent in all this is the growing schism between the established rancher and the upstart. Within the nature of the rancher, this schism has always existed; but today, the monetary policy widens the gap by crushing the beginner, where it once rewarded his courage and effort.

As I consider the established rancher, the established ranch and the established rancher's wife, those born and raised to the earth, secure in the productivity of their ranch and their relatively debt-free position, I find it is like the recognition of a different breed of person. The rancher's lifestyle and financial ease are taken for granted as only the rich can be sure. Though not with the same wealth of the very rich, within their echelon they feel the same warm security of certainty.

Whereas the upstarts, or those who are being crushed by the debts of unwise expansion during these past few years, are in contrast nervous, agitated and confused. Hard work and achievements that excel the performance of the established rancher, no longer have any guarantee for his success. The gambling odds have shifted upwards.

The cost of the cow is no longer the big cost of ranching. It is land, equipment and operating costs. Forty years ago the cost of a cow was about $40, and the calf returned about eight cents a pound. Today, that cow costs upward of $700, and her calf returns perhaps 70 cents a pound. This results in an even smaller ratio of return than it did 40 years ago, with infinitely higher costs.

It is true that today's performance is better, the number of calves available to be sold greater, but that cannot begin to offset $1,000 a month ranch hands against $30 a month cowboys of $1,000 an acre land against $30 an acre, or financing at 20 percent against 3 percent.

One can't help but question what percentage of ranchers who have bought during this past decade are still ranching today. The upstart is nervous, agitated, and confused. How can he possibly feel different? His ranch may be as large as that of the debt free rancher, his gross earning as high, but the banker and the government confiscate his living and his profit. He has only this against him, that he came along too late.

NINE

FURTHER EXPANSIONS

Every person in real life seeks, above all, to get some control of his income. It is, in fact the most sought after and cherished of liberties.

JOHN GALBRAITH

FOR THE CATTLE producer, there was nothing unique in the low cattle prices which existed from early 1974 to well into 1978. These high-low cycles were with the primary producer from the beginning of time. Higher cattle prices of the early 1970's, low grain prices, and the federal government's generous encouragement with guaranteed cattle purchasing loans had accelerated herd expansion. Heifers were bought and held for two years to become breeding cows. Transient livestock producers, mainly within the grain and corn producing sector, decided this was the time to make extra dollars by producing or feeding cattle. With the surplus of livestock held back in anticipation of even higher prices and the inevitable consumer resistance to rapidly increasing beef prices, the days of high prices were numbered.

What was, however, unusual about this trough in the price cycle was the unprecedented length of it, coupled with the newly-felt effects of inflation. This added a new burden to the singular, rooted rancher.

During this slump, the B.C. government stepped in with an Income Assurance program for B.C. cattlemen. Since the Agricultural Land Reserve Act of 1972 had taken away from producers their right to subdivide their high cost properties, the government felt some obligation to ease their plight.

In our particular case, the program was of little benefit. Our objective was to build up our cow numbers, in part by buying good cows as they became available at dispersal, but primarily by holding back 80 percent of our heifer calves. As the income assurance was based on calves, or yearling pounds sold, our share was significantly lower than it would have been had we chosen to sell our young stock rather than improve and increase our herd.

Ranch timber sales continued to offset a portion of our operating and development costs, but there seemed no way that we could inhibit our rising debt load. Unless we were prepared to sell off some of our non-arable land, our attempt at economic sufficiency would be shattered.

We resisted losing the unique quality of our ranch by the addition of proximate neighbours. The problems of loose dogs and carelessness with gate closing would rapidly increase. From the positive side, we were aware of the assistance good neighbours could bring, and the greater chance there would be to obtain telephone service. All these arguments, however, were really an exercise within a choice we did not have. We could see down the road very real cash flow problems. Choices really consisted of these options: selling the ranch, a quarter section, or sub-divided land. We made our decision. Two lots of class 6 and 7 land, each ten acres, would go a long way to lowering our debt load and the interest drain.

In the fall of 1976, Henry drew up and submitted the subdivision plan. The highway department assured us that the preliminary approval would take about two months and final approvals, we calculated, would be an additional three months. This should mean

Henry, Elizabeth, Eric, Irish Wolfhounds, 1972

Eric training a bull for show circuit

Full blood Arab mare 'Somenko' autumn slopes

Henry, Angus, ear tag a calf

that the sale funds could be counted on by the early spring of 1977.

Had we had the time to pay attention, we would have recognized that this was an absurd time estimate. Wishful thinking it was, but not based on the facts of the state of subdividing in B.C. Any land related in any way to the Agricultural Land Reserve - in this instance, our subdivision was part of a 320 acre parcel, 100 acres of which was in the reserve - would have to be cleared by the Land Commission. They met only monthly, therefore the backlog even after over three years of operation would be cumbersome.

We prodded and nagged, riding herd on the Highways Ministry and the Land Commission by means of our radio phone or by pay phone. Finally the reply from the Land Commission office was returned, not to us, but to the Highways office in early spring. The Land Commission had unequivocally rejected the subdivision but chose not to give a reason why.

Angered by the high-handedness of these dictates, I immediately placed a call to the Administrative Assistant who had written the letter. Without hesitation she offered *her* reason for rejection, which was, that we should sell a 160 acre parcel instead of the two ten acre parcels. To this, following my questioning, she added that by going this route someone else would have sufficient land for an economically viable operation. I was perplexed by this conclusion.

"Do you mean that it is your advice that we weaken the grazing capacity of our ranch by the sale of 160 acres of grazing land, rather than sell 20 acres of hillside rock?"

"Yes, that is the best course of action."

The certainty of right in this woman's voice, the new implicit power she held at the end of the phone made me want to yell at her, "Dumb! Dumb! Dumb!" Instead I heard myself say calmly if impatiently, "Are you aware that it takes 35 acres of grazing land here to graze a cow? Do you know that there is no water available for those 160 acres; that the irrigation rights on the creek are fully committed? Do you know the hills, the ravines, rock out-cropping of much of our land? How can you speak of a viable operation on 160 acres here? This is not the highly productive Fraser Valley!"

"That is the decision," was her equally impatient reply.

I drove back to the ranch to share with Henry the reasons for the refusal. He was angered at the inadequate data that the commission was basing its decisions upon. Helter skelter, they were deciding the fate of human effort without the geographical knowledge or consultant's advice to guide them.

Henry did not wait for a reply to his scathing letter of protest. In person he went to Vancouver to confront a very able, but obviously ill-advised, overworked chairman of the commission. When the total picture was learned, the options *really* analyzed, a reversal of the commission's decision was instantaneous.

We had, however, lost precious months of heavy interest-paying time. It would be early fall before the final release signatures were obtained and the subdivision duly registered. Through summer, delays were encountered again and again by the absence of officials on month long holidays. The bedraggled plans shuffled around the country from Ottawa to Victoria as they awaited well-tanned officials' return for signature. By March 1978, exactly 18 months following the submission of the subdivision plan, the lots had fine new owners, and we deposited our shrunken cheques in the bank. At this, I couldn't resist commenting to Henry, "And to think we moved to the country to avoid bureaucratic entanglements. My God, we spend half our working time dealing with them!"

With the funds from the lot sales and an additional Farm Credit development mortgage obtained in early 1978, that year should have been a financially unpressured year. I don't recall it as such.

Much of the funds from the lot sales, shrunken by inflation and their delayed arrival, went to pay some unpaid commitments from 1977; the rest carried us through the commitments of 1978. Between 1972 and 1977 our total yearly income had been only $115,000, yet we had made improvements and herd cost expansion of that amount. In addition, we had operating costs of $95,000 and payments of principle and interest of over $30,000 for a total shortfall of $125,000. Had we known what was yet to come, we would have wisely sold the ranch then and there.

But through those years it was often Eric's youthful optimism and belief in what we were doing that encouraged us to carry on.

It was the beginning of unpredictable, ever floating upward interest rates that belittled even our most careful financial projections. The small margin of net income we saw on the horizon was as quickly swept away by a rise in interest rates by as little as one or two percent.

As part of our development plan of 1975, 80 more acres of land were to be cleared in 1977. As in the past, two big D8 cats and their owners arrived to stay at the ranch, with family to work the better part of three weeks on the first stage, rough clearing.

These operators, a father and son team, were skilled and careful land clearers. With the use of brush blades, there was very little of the valuable topsoil removed. The following winter, the windrows and root piles containing little soil, thus burned relatively easily.

This land that had been previously logged was to cost us $150 an acre to clear. Another $200 would be needed to root and rock pick, level, prepare the seed bed and seed the acreage to alfalfa. In addition, $250 per acre would be needed for extension of the mainline irrigation, for sprinkler lines and irrigation guns. With the total costs of $600 (excluding land purchase) and with valley land selling at $4000 an acre and up, this seemed a sensible investment. Of course, most valley acreage was producing fruit at a yearly gross of over $3000 per acre, and our acreage in alfalfa could produce only 4 to 5 tons, a gross of $400. Our land could produce the hard fruits, even at a 3200 foot elevation, but in 1977 B.C. needed apple orchard expansion about as much as North America needed more beef production.

Assuring ourselves, however, that the world glut of beef would soon be consumed, we concentrated on alfalfa. We had chosen to be singularly dedicated ranchers, not mixed farmers. Although our herd of one 100 cows could not make us any tangible dollars for our efforts, the challenge in our undertaking was strong enough to carry us along. We continued to fence and cross fence each area of the hayfield as it was completed, and awaited and readied ourselves for that unknown day when prices would swing up and a herd increase would be feasible.

Each field, as an entity, had its name. This naming was needed, not only to identify and compare yearly yields, but as a direction for communication. Names like Gopher City, Grey Slope, Jaffe Draw, the Saddle, North Spring, Platter, and the Calf Meadow came naturally and far more readily as distinguishing names, than meadows one, two, three, four...

The calf sales that fall of 1978 finally brightened the eyes and gladdened the hearts of the area ranchers. Their endurance again had paid off as they watched their calves bring up to 75 cents a pound, a leap in price of some 35 cents. It all looked downright gluttonous until one remembered that the price in 1973 had been over 60 cents. But many ranchers would not remember. They were there to enjoy, even if briefly, their work at last rewarded.

Personally, we had learned to eye the increase with skeptical joy. We knew that unless the price continued to rise for at least five years, no recouping of the previous losses would occur. With learned suspicion of bank interest rates and growing inflation, we contained ourselves from this infectious sense of well being. Nevertheless, we were optimistic enough to consider herd expansion.

That fall we began negotiations with the rancher who held the main grazing rights to the 60,000 acre range adjacent to the ranch. He was anxious to cut back his herd, was willing to relinquish the range, and sell us a 100 of his better Hereford cows and the expected, but not implicit, rights to the range. For this gamble of obtaining the range, we were willing to pay a premium for his bred cows.

Cattle with range familiarity are nearly essential to an easeful operation, and as these cattle knew the range, we decided to proceed.

In this case, scarcely had the ink dried on our second mortgage when we were back again to refinance a third short term. This mortgage would cover about half the cost of the cattle purchase with additional operating loans obtained for wintering hay supplies. As our hay production was not yet able to cover the feeding of these additional cattle, a loan of more that $30,000 would be included in the operating loan for their feed supplies.

A costly gamble, perhaps, but without this grazing and cattle expansion, our ranching days were numbered, unless we wished to subdivide indefinitely. If we could not rid ourselves of inflation, we had to go with it, expanding faster than it increased. As much of this expansion would be financed by the 29 year mortgage, not by the usual five year bank loan, our annual payments would increase only marginally in proportion to the projected increase in income. Within a year, at the most two years, we expected our hay production to be adequate for the complete herd.

With every step, however, that we tried to take forward, we seemed to see our way thwarted by changes in the rules of the game made by government officials. Obtaining the range grazing rights would be no exception to this pattern.

Historically, hay production commensurate with the numbers of cattle to be grazed was the prime requirement for one to be considered for range grazing rights. In fact, in our earlier contacts with the range division regarding a range permit extension, they had adamantly insisted that, without the development of our hayfields, we had no priority except that of our proximity to the range.

In spite of our growing hayfield acreage with yields approaching 400 tons by 1979, and in spite of our ranch position surrounded as it was by the range, and our purchase of the range cows, we soon were to learn that we had no special position as a range applicant.

A new Range Act and a different decision-maker, who did not believe in the need for commensurability, and believed instead in breaking down large ranges to cooperative units, was in the driver's seat. All the previous established priorities were being abandoned.

The intent of this new local system was to increase the number of permittees per range; this was aimed at encouraging the sharing of responsibilities for fence repairs and for cattle movement.

In theory it seemed feasible, but in the years of history, it had shown that there were always one or two of a shared range group who did all the work. There were always the skimpers with salt distribution, the ones too distant to be available for a range emergency, the one who over-grazed or over-stocked, or did not remove his cattle as directed.

A shared range was not the route we envisioned for our needs. With our selective breeding program, our top bulls and our wish to use range mineral blocks, we could not reconcile fitting this in with other less dedicated ranchers' views.

When the decision came down, we learned that the range would be shared. A long-time resident, an orchardist who raised cattle, but harvested not one ton of cattle feed would be the other permittee. In spite of our protest, our working through the appeal procedure available to us, we were stuck with this somewhat unworkable situation. One concession, however, was given to us. We would be given, for our sole use, a half dozen spring breeding pastures; the summer would mean a mixing of cattle and bulls. Even though it seemed sensible to divide in two the summer range with the other tenant using the south end closer to his home headquarters, and us using the north, we were both in agreement with this idea and were prepared to pay the costs of the cross fencing. The request for this workable solution was denied.

We had one advantage here. The fall range located in the north reaches of the range was at the opposite end to the other rancher's headquarters. His cattle would have no homing instinct to drift to our direction. Short of a round up, and a long cattle drive during their busy apple picking time, few of the cattle would ever arrive at that range. The disadvantages of this whole scheme was that unless the other rancher removed his cattle from the summer range at the allotted deadline, there would be serious overgazing of that area. In future years we would both suffer from this abuse and overuse.

Though we were less than happy with the range setup, we would need to work within it, and within the previously committed initial undertakings of the Agriculture Rural Development Subsiduary Agreement (ARDSA). At that time this was a government/rancher funded range improvement scheme. Although Henry's, as well as other ranchers' ideas and objectives would be requested on all changes and range improvements, they would rarely be taken into consideration.

Fence building that Henry deemed to have priority would be done at the end of the program; thinning of trees which he pointed our as being advantageous to both timber growth and grass encouragement would be ignored altogether. Instead, top priority was given to the replacement of the nearly inaccessible Indian Reserve boundary fence. To cause us more annoyance, this border fence we had laboriously put in, at our own expense, just some four years earlier. A subsequent Indian Reserve boundary survey had since indicated that the fence was up to ten feet within our side of the line, and because of the Indians' strong need to define exactly their Reserve boundary, the fence had to be replaced. This was done, not as we had done it, with the use of pack horses, but with the costly use of a helicopter to fly materials to sites along the route. A better fence, but a lot of taxpayer dollars. In return we were left with the time consuming job of removing and rolling up our old barbed wire from the dangerous narrow, lock-in passageway formed by the new and the original fence.

Whatever our dissatisfactions, however, we had our range and could now run, with the use of our deeded land, up to 250 cows and winter with our increasing hay production up to 400 head. We, at last, could see ourselves as ranchers.

TEN

THE HIRED HAND

The man of stamina stays with the root
Below the tapering
Stays with the fruit
Beyond the flowering
He has his NO, and he has his YES

LAOTZU

FINDING, HIRING AND holding good ranch workers is never easy. Most young people imagine ranch life as a pastime of riding the windswept ranges on a favourite horse, roping calves and playing cowboy. Perhaps some of them are aware that there could be fences to repair and corral rails to replace, but few of them know that these jobs are far more than a small part of ranch life.

Most seemed unaware that the main jobs of summer were haying and sprinkler changing and, at least on our developing ranch, rock picking, and new fence construction.

Some of the easily discouraged fellows, finding feed buckets in their hands instead of reins, shovels instead of ropes, quit by the early hours of the first day. Others moved swiftly over the hayfield slopes, casually balancing 30 foot irrigation pipes through the cool May mornings of their first day on the job, only to grind down to a crawl, develop back problems, or stomach aches when the hot morning suns began to shine.

The special worker often geared himself to ranch routines and remained with us long enough to handle the various sprinkler systems, to learn the content and boundaries of the ranch and range, to learn finally to use a shovel, a hammer, a chain saw and then quit, forcing us to scurry around to find a summer replacement. Both Henry and Eric bemoaned the continuous time spent and lost in training and supervising.

As part of an attempt to provide employment for the young unemployed and students, our provincial government offered a 50 percent salary rebate toward the total wage of qualifying youths. On the one hand, this program allowed us to pay a higher wage, but offsetting this was the amount of additional paper work involved, and the quality of persons available.

This sharing of costs tempted us to hire the young and not investigate the more mature workers—workers we knew to be more stable and better equipped to handle the heavier, more physical work of ranching. The discouraging truth regarding most of these programs is that the work output of two young people is usually equivalent to that of one more mature person, but has the added disadvantage of requiring the supervised training of two personalities with their concomitant frustrations.

In the beginning years, ranch accommodation was always a problem, thus we were narrowed into hiring mainly local boys, who usually joined us as graduates of routine, but socially oriented, fruit-picking summers. Their anticipation of a summer job, more challenging and manly was often quickly dissipated.

Over the early years we had them all. The overweight, easily tuckered out who quickly left, offering the old and tried reasons for sudden departure:

"That job I applied for, before I started here, they have offered me work in shipping now."

Of course, it had never been mentioned that he had an application in elsewhere, and would leave us if he were called.

Then, there was the wiry, over-enthusiastic work zealot, who insisted with a religious fervour that he wanted work, wanted a ranch job, and sold us further with his lengthy tales of the demanding work that he had done on prairie farms. But Monday came and he did not. A week later we would hear from him, with excuses of a family death and his too abrupt departure for the funeral to phone.

Then his pleading question, "Is the job still open, can I have it?" Knowing better, we responded to the easier route of giving him a second opportunity to prove himself. "All right, be here at eight tomorrow." A couple of dedicated days of work and the young lad would be missing again. This time, the evening call would excuse himself for reasons of his grandmother. "You see, I had to take her to this fair. She loves these fairs."

There were the body beautiful boys, who spent their spare minutes flexing their muscles, their first evening chain sawing and installing supports for a punching bag, then, in spare time, working out, eager and flamboyant in the limelight until the passing days no longer brought an audience.

One of our early summers, in an attempt to relieve Eric and Henry from the constant evening sprinkler changes, I gave in and allowed two hired local boys to spend their summer with us. Although their accommodation would be a tent-trailer, they would have some use of the house and join us for meals. To offset their evening sprinkler changes, they would have free hours during the afternoon.

Within days of this arrangement, I knew that this would be one of my longest, hottest summers. Essential as these boys were to us, I found myself confused and angry at these 18 year old put-downers.

At meals, Jack, a garrulous 'know it all' bore, showed manners which would match only that of a chimpanzee. Regardless of what delectable meal I placed before him, he showed a rudeness to his eating—picking and abandoning morsel after morsel. Not hungry, he was to return within the hour for a banana, or to take time to gulp down a bowl of granola. The clutter of his sprays and toiletries in the bathroom was only matched by that of the most meticulously groomed doll.

Unfortunately, his team mate, Ron, a more moderate lad, in proximity soon took on much of the colour of the lowest behaviour of Jack.

I said that ranchers were long-enduring. We were. We thought that the boys, now used to our different manners, and environment would change, but neither their behaviour nor attitude did.

By summer's end which for me had seemed to be a redundance of hauling food in, cooking for the four insatiable men, doing dishes and garbage runs—we had learned two things. At least, in our intimate quarters never to offer board to more than one worker, and secondly, never to hire buddies.

Although we had been approached for work by ranch-trainee students, Henry soon discouraged them from what they thought would be a horse riding summer. But, with our son away at college for the calving season of 1981, we decided to hire a capable young girl whose experience, like most female applicants, was more with horses than with cattle.

She soon proved herself patient with cattle and calving; skilled with the occasional mothering up of heifers and their calves, and as capable as many men at using a tractor and loader. Our awakening here was her attitude toward the time clock. Experienced as a nine to five worker, she automatically assumed that when the clock struck five that she would pack it up, even within an incomplete job of cattle moving. She was slow to learn that ranches really are different from offices and stores, and that differences exist in far greater terms than those of working inside, versus working outside.

Aside from these transient workers, saving us from complete despair, there was always Angus. Every ranch needs an Angus, but not every ranch is fortunate enough to have one.

Angus had come to us in 1976 as a part year employee. He was a well-seasoned cowboy who had just left the employ of a ranch which had been broken up and sold. Raised in this area, easily using directive names such as Benny's Flats, Indian Meadow, Bohunk Meadow, Walker Corrals, the Sidley Cabin, as if they were the first words that he had ever spoken, he knew the range better than we would know it for many years. A prince of a man, he had all those rare old qualities like punctuality, reliability, steadfastness, an ox's stamina and the tenacity to see a job done, no matter how trying, or what time the clock read. He was a blessing we could not easily part with, and in spite of what our cash flow indicated we had him working full time by 1977.

It would have been more advantageous to have someone with us, who had mechanical abilities as well, but we sacrificed these skills for those essential qualities Angus had to offer.

Angus became the embodiment of the antithesis of Henry. Mounting his horse, Henry lived his life riding off in all directions. Angus had the steady, slow pace to his life of a long trail walker. I don't think he knew what the words 'rush' or 'urgent' meant. Not with a "Hi Ho Silver,' but with an easy walking gait, he and his horse would saunter off into the noonday sun for a casual range check. His pace would remain that way unless he encountered a problem. If he did, then he would rise to the most demanding occasion and manage with his years of uncanny horse and cattle sense to solve the problem. Sometimes, it was a fence repair, sometimes he would bring an animal home, often for miles, for care and doctoring. Even a bull he managed to remove from a group of cows, proudly to appear at the gate with it.

It was a couple of summers back when Angus proved, again how very valuable he was to us. It had been reported to us that a heifer calf had been seen up on the range with a snare caught tightly around the ankle bone of its front leg. The calf apparently had stumbled upon the one snare accidentally left behind by our local winter trapper. Unless this calf was quickly located and the snare cut off the ankle, the calf would soon lose its leg, if not its life. Several trips out with the pickup truck finally located the four month old calf which was in pain and very lame from the swollen leg, but was still very mobile on its other three legs; too mobile, in fact, within the thick brush to rope and doctor there.

Angus, as the only person free to go, and the best person to go, was sent off on his horse to bring the limping calf, and mother back to the ranch. This cowboy had the patience and quiet manner to move the two gently and slowly home, inflicting as he would as little stress or pain on the calf as possible.

As was expected, within a couple of hours Angus came through; the calf was swiftly immobilized in the calf tipping table and the snare, biting into the bone, was quickly cut off and the wound sprayed with antiseptic. Within a week the calf's use of the leg was nearly back to normal.

Angus's cow sense in riding range, and fence building were his specialties; thus we encouraged him to do mostly these jobs even though Henry or Eric could have benefited at times from a summer saunter through the cool upland range.

Occasionally, during the early years Henry privately became impatient with Angus's slow, unyielding manner. Later, he learned to accept Angus for what he was, a man past changing, and beyond worrying. We learned to see him as relic of the past, part of a history now gone, having a special place we could never enter.

We had with us during two calving seasons some of he apprentice students of the three month long Okanagan College ranch-assistant training program.

This program had finally analyzed the ranchers' needs and attempted to answer them. Not solely a make-work program or a specialty training program, the college in this instance has provided training for a much needed British Columbia worker, the present day's version of the cowboy.

The student was to arrive at the ranch for a three week stay, with two weeks of comprehensive instruction and field trips behind him. He, or she would bring with him at least some background of the ranching terminology - know a Hereford from a heifer, a balling gun from a syringe, the appearance of grain from the appearance of fertilizer, the symptoms of shipping fever from those of coccidiosis, and had had some training in ranch methods and ranch practices.

Perhaps it is was the co-ordinator of the course, a rancher himself, who sent them out with realistic, not imagined, expectations of ranch life. They arrived with one thing in common; the fervour to learn and the flexibility and wish to fit into whatever routines, hours and living conditions befell them.

And what a breath of spring their eagerness could carry into the late winter spirits of the rancher. How exceptionally well-timed, this fresh energy has been planted. By late January the rancher, fatigued from a long winter of feeding and ice chipping is already questioning where his energy and dedication will come from to proceed into months of calving. Then the student arrives, occasionally a female, in the spiff of cleanliness and eagerness like a beneficial new doctor joining a M.A.S.H. unit.

Though these young people differed as to background, experience, and education, and their reasons for taking the course were varied, I sensed my relief when they knocked at the door, knowing that for three weeks our hands would not be quite as full.

I don't wish to imply that the track record of these students was vastly different from the young summer employees. We still lost hammers, had gas caps left off trucks, had shovels cast aside and forgotten, gates left open, our combs and brushes treated as theirs. But what made them different from the job-oriented summer help was their attitude, their sincere wish to learn and to tackle any new job or problem placed before them.

These trainees had willingly put themselves in a non-pay learning experience; whether they would continue in this type of work was not the issue. The issue was to learn from whatever that situation has to give. As is the earlier, unpaid nurse training, the basic soldier training, the Peace Corps, Katimavik, or CUSO, they were learning and growing from experiences of the real world. Regardless of the work direction they chose, this experience would carry them forward. As Masanobu Fukuoka recognized so well, "The ultimate goal of farming is not the growing of crops, but the cultivation and perfection of human beings."

Perhaps a few of these young people learned to understand some of that truth. They would learn some of the skills of rural living—to use their backs, hands and heads—to be attentive and aware. All they had to do was to assist at night once, with the round up of cattle on the loose, and they would have learned the necessity of closing gates.

For years, ranches have been built and maintained by the skills and teaching that was provided by the rancher to his unpaid sons and daughters and to his unpaid wife. It is unfair to expect that he must continue to teach, instead, to other men's sons and daughters the generic skills of living, in order that the price of food continue to remain low.

With the ease of city living, it must be important to ensure that the younger generations learn other living skills than the chore of taking out the garbage. In the past, the chopping of wood, the use of a saw, a hammer and a shovel were learned in the home. Today these frequently go unlearned, as well as the ordinary adaptive knowledge accompanying them.

The rancher may still be seen in the role of life's instructor to these youth. This is well and good provided that the rancher has the energy, time and patience to teach others as well as to benefit himself from this experience.

The root of the problem of the hired ranch worker, however, lies outside the training programs. Today, as in other segments of our society, it lies in hourly wages and fringe benefits.

As long as ranchers are denied a way to pay a wage nearly equivalent to the industrial wage, the number of qualified workers will not be available.

If semi-skilled workers are guaranteed a living wage from the market place, why should a ranch worker be treated differently? Why should ranches and farms be denied good mechanical workers while these unemployed workers receive as good, or a better income from unemployment insurance? Why should the rancher, because he cannot pay a competitive wage, feel that he is a second class employer? Why must he continue to fall back, for the bulk of his help, on the training program employees?

A discrepancy lies between the farming sector, and the industrial, technological world. This has had a legitimate place. Can it any longer?

ELEVEN

THE MAN

Each one of us who cares doing his small, whatever-he-can-do thing in one last fight to save the land. I must not let him down.

DAYTON O. HYDE

BY THE TIME my husband had reached his fiftieth birthday, by most people's standards, he had fulfilled his life. As a college student, he was a skier, had raced and trained for the Olympics. As an architect, he had designed some of the coast's most creative and exciting buildings and homes. As a cattleman, he had built up the nucleus for a fine quality herd and a ranch which basically had little to flaw it. Other cattlemen had begun to respond to, and respect his growing knowledge and dedication, his single-purpose direction.

It was never Henry's nature to turn away from more efficient systems, whether they were feeding methods, better equipment or tools, alternate farming practices such as silage making to counteract the weather, or a special feed wagon to take out our supply of apple pulp which for many years, we hauled in daily and mixed with grain or supplement.

As keeper of the purse, no matter how often I pointed out our financial shortages, he as quickly forgot, or ignored my warnings and proceeded with his carefully researched expenditures.

A ranch as cumbersome as ours, on grounds of better efficiency, could justify for the next 20 years any purchase, be it cattle fountain waterers as insurance against a deep creek freeze, a calf tipping table to assist with calf doctoring and branding, sprinkler guns rather than sprinkler heads to decrease the time of irrigation changes, a four-wheel drive rather than two-wheel drive truck to negotiate slippery winter hills more easily along the feeding route. With labour costs escalating, even though we could not afford the high interest and short-term repayment of capital purchases, neither could we afford not to have them.

It was not that purchases were foolish or decided upon rashly. On the contrary, Henry investigated expenditures carefully and rarely new, usually used equipment was purchased, then well maintained and serviced as our growing repair bills indicated.

Like blossoms in spring winds, cash came in and as quickly went out. Too frequently our operating loan soared because of unbudgeted items. Always, we were under pressure of cash flow, delaying a payment to Peter to pay Paul. The bank continued to support us, or was it that we continued to support them? Still, Henry resisted and resented the thwarting of his land development plans by the unprecedented high interest rates and high inflation which continued during those early 1980s.

It was apparent that land development no longer penciled out at 20% on borrowed capital. Further land development would be detrimental to our already strained payment schedule. The demand for Okanagan Valley lands had increased our land values ten times in as many years, thus our debt/asset ratio remained healthy. We could assure ourselves that our cattle inventory could cover most of our debt load. But neither of these conclusions helped us to proceed, nor dimmed our awareness of our precarious cash flow position.

In fact, Henry concluded, as an example of the bleak picture, that an inherited 250 head fully-equipped debt free ranch could not generate sufficient income from the calves of those 250 head of cattle to repay the bank loan for even the cattle purchases over that usual five year period. These were indeed dark conclusions, at that time of the country's economic plight.

Many ranchers cannot help but agree with the old farmer's reasoning, "I wish these experts would stop telling me how to farm better. I already know how to farm better than I can afford to."

But Henry still refused to do less than he knew how to do. Of course there was nothing ordinary about Henry. As much as I saw myself as a person who primarily responded to the deeper needs of human beings, Henry was primarily driven by his need for precision. Having an indefatigable ability to grasp and centre on a direction, he never allowed an obstacle to deter him. He rose earlier, slept less, demanded equally of himself, as well as others, was patient or impatient as the situation warranted. His enthusiasm was never clouded by consequences, thus he refused to ranch by the 'seat of his pants'. Instead, as ranchers have a way of doing, he buried himself in what he wanted to do, ignoring as often as he could, Canada's and our own economic gloom.

Because of cattle handling costs and management, not because it is the better way, most purebred/commercial ranchers separate these two herds from each other, babying the pure-bred herd while expecting the commercial herd to fend more for itself.

Our complete herd, commercial and purebred, was treated as one unit with the recording of performance data on every animal: birth weight, tattooing, ear tagging, identifying of sires, weaning and yearling weights. No special favours were granted the purebreds in care or in culling. Only the sick and poorly-doing animals in the commercial or purebred herds were given special treatment. A back-up milk cow that some commercial and purebred breeders relied on was never used, as this use could alter our true performance figures.

Many less dedicated ranchers would have stared in disbelief at seeing our early attempts, with inadequate corrals, to sort cow-calf pairs into groups for spring pasture turn-out. This required the better part of a day of utter turmoil to sort, cut out, match up cow with calf before a group could be readied to move out, this not counting the previous day's vaccinating, and branding. The subsequent day, that group would be corralled, and later driven to their breeding range. Their range could be as far as two miles from the corrals, but the procedure would continue daily until all groups were cut out, and moved to their spring breeding areas.

Later, Henry was to make two significant changes to streamline the operation. The first was to ear tag the calves at birth with the mothers' herd numbers rather than the consecutive birth numbering, as previously done. This facilitated cow/calf match-ups. Later, the calf would receive a second tag bearing its own birth and year number. However, these changes were more important in that it created cow groups through the calving season. Cows or heifers and their calves, as they calved or soon after, were moved to one of seven allocated ranch feeding areas. These groups were formed from previously selected females that were intended to be placed with a specific bull. Each group could then be run separately into the corrals for re-checking, vaccinating, branding and thence as a group moved to its breeding pasture. This meant more work during calving, but did scatter the herd and thus reduce the chance of disease spreading.

To operate a large herd in every way as the better purebred herds are operated is costly in terms of labour and time. Without the advantage of at least 75 percent purebred prices, even in good times such a practice is financially impossible. During the later years, our numbers were such that we could sell off up to 20 percent of our stock at purebred prices, but we were still not sufficiently satisfied with our product to begin to market purebreds in an intensive way.

It was still difficult for range growth and performance to match most breeders' lush pasture growing and their controlled environment; but Henry was not inclined, as yet, to alter this true, homogenous testing program. If we could not produce the biggest cows, he could produce the soundest ones.

Within our management capabilities, we strove for better and better cattle nutrition and care. Wintering cows were not over fed, but were adequately fed to produce healthy, vigorous, but not oversized calves, with probiotic mineral mixes in feeders. Wintering of calves and yearling bulls were given more selective rations. Mature bulls were fully supplemented with grain, were brought back to breeding condition readiness, but were not fattened. All cattle were fed healthy, organic feeds.

Even with poor cattle returns, Henry did everything to curtail the neglect of the cattle. Not that the ranch could run perfectly because ranches are chronically understaffed during pressure times or crisis. Although western ranches do have to their advantage the use of cheap summer grazing for the cattle, they soon lose this edge over the more functional pasture fed cattle farms because of their need for extra manpower to handle their cumbersome operations. Thus, even with being concerned, a too thin animal is often neglected too long before time can be found for diagnosis and special care. Manure and mud may remain in corrals and pens too long; groups stay in wet areas too long before time can be found to move them to total dryer lands.

It seemed our herd was always expanding faster than our facilities could be readied to accommodate them. Physically the body wearied before all that should be done was done; the stress felt at not doing what should be done was always a part of our lives.

Within these limitations, Henry continued his efforts to build toward a super herd. Even on the commercial herd, only the very best bulls were used. Yearly purchases were made of the highest performing Charolais polled bulls, with many of our females artificially inseminated.

Initially, we had calved on our pine covered dry south slopes. This excellent site soon had to be abandoned because of several pine needle abortions, and the loss of a cow from bloat. We were to discover that any ruminant can become easily trapped in calving position, in even a small depressed basin, and can die from the inability to expel rumen gases. In spite of the fine protection the trees and early sun gave to the cows, and the hillside exercise which assisted their calving, we had to move the calving to a flat portion of the hayfield. The cows were more exposed there, but their visibility made checking easier. We later built cubicle calving sheds where heifers and problem calvers could be dealt with in warm, well-lit quarters.

As we started calving earlier in the year, with that to begin the middle of January, these sheds were used more extensively. Cow and calf were given 24 hour protection following calving, before they were turned out to the dry hillsides. Eric helped streamline the operation by designing and building a four by four-foot incubator equipped with a heat lamp. An hour or two within this warm interior was to save us from losing many a chilled, exposed calf.

In 1980, we purchased one quarter interest in, and the physical possession, of an all breeds World Record Gainer, Canadian Meridian 12J, more commonly known as 'Big Guy.' This bull was jointly owned with the breeder, Dr. Fred Day, and New Zealand Sire Services, who had rights to market semen from him in New Zealand and Australia. Outside the fact that the bull had been a pampered baby and caused no end of grief during his first year on range with us, his giant 3000 pound frame managed still to produce 100 plus pound calves, which slipped out from the cows, even smaller cross-bred cows, bright-eyed, fast and without effort.

Bulls that have been raised in confined quarters can initially create very real problems for the rancher. Pampered as 'Big Guy' had been, raised in a small pen, he was not used to the jostling and fighting of other bulls, thus he had no way of knowing the full range of his physical movements. He had little, or no experience with rough terrain, consequently not only were his feet soft, but he didn't know where it was safe or unsafe to place them. Before he even reached the range, a fight with one of our bulls resulted in a hoof puncture and an infection.

We rountinely treated the infection and he was turned out to our neighbour's private range. There he encountered his very first horse, and in the ensuing battle, the horse did what he was used to doing, he kicked. The Big Guy was left with a severe hip hematoma.

We turned to the sulfa drugs to treat his ailment, as well as the redevelopment of the foot problem. We were unaware of the sometimes suspected claim that some bulls are allergic to these drugs, and that sterility could result. Whether the rash and tremours that we witnessed were symptoms of an allergic reaction to the drugs, or whether it was his weakened condition that caused the sterility, he was tested sterile. Although he appeared to be breeding the females that were later placed with him, he had, in fact, not bred even one.

It was not until August when we sent him to an Artificial Insemination Centre in Alberta in order to have semen drawn for our New Zealand partners that we learned of this sterility. The A.I. veterinarian suggested that we should not give up completely on this valuable bull, efforts should be made to routinely test his semen over a period of time. Prior to Christmas there were no semen improvements, and reluctantly, we all decided to end his breeding career. Because of the season, we decided to wait until after Christmas to allow him one more test. In this test, like the miracle of Christmas, his semen returned to being completely normal.

How close we had come to losing the significant impact that this bull would have on our herd. How near he had come to having no chance for champion descendants.

In addition to 'Big Guy', additional impressive other bulls joined our bull battery. If we could not buy super cows, we were going the best route we knew how to produce them.

Still, within all this progress, it was clear that we were, after all, only where we would have started from had this ranch not been abandoned; had this ranch been allowed to evolve naturally, had water for irrigation been developed 30 years ago, had fencing and cross fencing, land clearing been completed, had barns been kept and repaired rather than burned down, or dismantled and sold. If all this had been done before, not attempted at today's level of interest rates, how our ranching progress might have been accelerated.

If ranching in the past had been a more stable income producer, perhaps this ranch could have held its own, thus allowing Henry to do what he wanted to do, to ranch. Instead to try to keep ahead of inflation, he crammed developments into ten years, which in normal economic times would have taken 20 years. Then we had to face the unkind, often discouraging recognition that if we had done nothing to the land, its value would still be close to its ranching value today.

During the early years of quickening inflation, this attempt to out-run inflation seemed to be possible. Since 1975, however, in ranching there seemed no way to outpace the combined strangulation of inflation and interest. Too long continuing; if not soon to change, the last ranch indeed was being built.

TWELVE

THE NEXT GENERATION?

What other business can a man work at where his actions are supervised by a bunch of circling vultures just waiting for you to make a mistake?

DAYTON O. HYDE

ONE WONDERS, IN the midst of the turmoil of today's world with the growing uncertainty of employment, whether it is not best to deliberately try to curtail a son's inclination to pursue a broader education. That is, providing there is a source of employment at home—a solid business centered in the family—whether it is in the area of construction, retailing or of agriculture.

Should a son be encouraged to dream of unexplored territories in science, law, medicine or another vocation in the wide world of opportunity? Or should he be encouraged to follow family tradition, to plant his feet firmly in what he knows, to pivot around the home base learning only its specialties? A son's temperament and personality have a strong say in his suitability to go or to stay, yet as we all know, a great deal can be done to convince him where his place should be.

Eric's early dedication to the ranch, his involvement at the right age, the added advantage of isolation, and the immediate heavy work load, made him a natural to become a promising, second-generation rancher. It would have been very easy—had we had real confidence in our ability to develop an economically viable ranch within ever escalating costs—to have intensified his dedication to the ranch. Narrowing his boundaries, we likely could have succeeded in channeling his time, energy, and intellectual concerns exclusively into the extensive boundaries of the cow, the land, and the paper work of the operation. But never feeling secure in our own future here, or in the ongoing existence of southern B.C. ranches, and recognizing as we did the amorphous state of the livestock industry today, it was impossible to curtail Eric's interests or to limit his education, even if it had been our natures to do this.

His intense involvement in the ranch development did teach him the skills of life: the discipline of long hours of lonely, sometimes distasteful work, the need to act quickly under crisis, to understand the ramifications of financial choices. This experience, he knew, would give him a real advantage in the working world, and when it came to this working world, his choices and interests were broader than the confines of ranching.

The young today—educated as they have been by scientific models and systems analyses—have learned to favour expediency, and predictable quick profits. Their projected direction for the future does not easily accommodate patience, endurance or postponed rewards.

How much this education is in agreement with the more scientific farming of crops, which (except for the vagaries of weather) has a greater degree of predictability than exists in livestock production where the main predictability lies within oneself, one's dedication and management skills, and where losses must be tallied up as deaths. This constant reminder of failure, measured and calculated by the heartbreak of death, is antagonistic to the very nature of being young. This state of despair was so intensely expressed by the son of a rancher, when two of his best cows died simultaneously from pulmonary emphysema, "My God, how can you bear it? You try to make a living on 150 cows and your profit is lost by the death of just these two cows."

Many sons find that they must leave the ranch. Other sons leave for broader education and experience. Some still find easy enough reasons to return, at least to the viable family ranch. This is some reason to believe that, with time, our son will be one of them.

Ranching used to be viable. With a long established and realistic balance of cow costs to land costs, with a back up of growing sons to help shoulder the demands upon him, and supported by ranch-oriented cowboys who expected little more than a way of life, a determined rancher could assume that during his productive years a ranch could be established, that a relatively debt-free operation could be passed on to be further expanded by one or more ambitious sons. Although the calving percentage could be low in those times, and the poundage of beef sold could be unimpressive in comparison to today's production figures, the ranch, with few inputs, low taxes and single-figure bank interest could persist and, in good times, grow.

This tradition has disappeared as surely as sheep ranching has disappeared from the Canadian way of life. To understand why this tradition has apparently gone, we must compare the past conditions of ranching with the conditions present today.

As the various governments of the country vie for sovereignty over land and resources, the rancher who once had quiet possession of his lands is losing his own sovereign control of it. He still wishes to, but is unable, in the same way to control land now wanted for other uses, land which commands a higher and higher price tag, land that is now designated, or expected for recreation, wildlife protection, or urban industrial and residential expansion. As the rancher's land has increased in attraction and value, his control over it has passed from his hands.

Those dedicated sons who once stayed on ranches continued

in that tradition; today their counterparts see their choices more clearly. Now they can go in directions that seem to be more rewarding. When the calving night is cold, or the haying sun is hot, and when there no promises of financial reward for their efforts, how much easier is a 37 hour week of plant or office work, and erasing cares during the evenings and week-ends.

Educated as he is now in broader ways and having learned to play, the son recognizes the chronic understaffing of today's ranches, and shuns the heavy work load. He perceives the real disparity in both time and financial reward that exists between working in a business or in government and labouring on an inadequately staffed ranch.

How easy it is for us to recognize that during the February to June period, our ranch, if in line with current industrial labour practices or government departments, would require one ranch hand per shift of possibly twelve hours for calving, one health specialist for observing and doctoring the herd, two hired workers for feeding, one for fence repair before turn-out, one for fertilizing, harrowing, seeding and irrigation layout and repairs, two or three weekend relief workers, an over-seer manager, and an office worker to complete a staff of eleven or twelve.

In comparison, our staff, to handle the total herd, which in summers would rapidly approach 500 head, and the spring activities, consisted of two hired helpers, possibly for a short period a third, as well as Henry, and my contribution. Under these conditions, 60 hour weeks are commonplace.

An additional reason for the demise of the traditional ranch is that the cowboy as he was known is no more. Even if he wants to be one, he cannot, for government policies have dictated otherwise. Wrongly so, perhaps, for these protective employee policies have denied those who wish it, the right to be totally involved in an enterprise, to belong there.

Many cowboys are truly happy as a monthly paid ranch worker with a place to call home. They prefer flexible off hours, to experience totally the work seasons of a ranch, rather than be a detached

37 hour a week disinterested worker. These human rights policies deny him the experience of working with a ranch, instead he must work for a ranch and, thus, his sense of belonging, of having an enterprise to share, diminishes. He is alienated and adrift, directed to be the same as a time clock worker, when he is not by nature the same.

With the government's desire to regulate work equality, the B.C. government, as of 1981, compelled farmers to pay both permanent and temporary workers, following 5 days of employment, statutory holiday and vacation pay. To many ranchers who rarely have a vacation, or statutory holidays off, let alone financial reward for such holidays, it must seem productively impracticable and excessive during these financially troubled times, even though he wishes for a fair deal for his employees.

Ironically, by their striving for excellence, ranchers also have lost control. In the past, large ranches continued operation despite frequent neglect of livestock; such an attitude today cannot be tolerated. Without antibiotics, animal supervision, or handling, the cattle reproduced in a manner similar to moose in the wild. As the rancher's knowledge of care and veterinary practices increased, so did his costs for medications and vaccines, for better animal nutrition, and thus for mineral mixes and fertilizers.

As he recorded and improved his performance, his costs for ear tags, tattoo devices, rumen magnets, calf and cattle scales increased, as did the need for additional trained labour to complete new tasks.

With herd care and performance taking on a vastly different value, ranchers now generally wish to produce the best product. One concludes that they are not dissimilar in attitude from the highly efficient Okanagan Valley apple producers. Few, if any, apple producers here would choose to have an untended, wild-appearing orchard that produced solely cull apples for the juice market. Yet, as enough of these orchards at this time, do not exist here, the juice processor, Sun-Rype must import shiploads of apple juice concentrate from foreign countries to support its production demands.

Like the apple producer, the rancher could not return to an unsupervised cattle herd of unkempt animals, though if he could, his savings could be immense.

If we could cut our wage bill in half, nearly eliminate our veterinary bill, bull purchases, and necessary travel costs to purchase these bulls, registrations and paper work, mineral mixes, grain, bull insurance, fertilizer and miscellaneous other handling costs, it could cut our operating costs almost in half. Would it really reduce our poundage marketed by a proportionate amount? Not likely, at least for several years.

Traditionally, without that goal of maximum production, the rancher was more effective in controlling his costs. Now with today's inflation, the high cost of operating, and with most of the rancher's needs supplied by unionized labour, the high cost of operating places a new strain on the rancher's resources.

Even though 35% of the country's economic activity is estimated to be generated from agriculture-related industries, the related industries, who were once on the side of the producer have aligned themselves with large companies, multi-nationals, and unions, leaving that 4% of the primary agricultural producers to sink or swim or be taken over. The individual agricultural producer is very much alone.

As the noted economist Galbraith has pointed out, an inequity at present exists between controlled or protected, and non-controlled or non-protected segments of society. Ranchers and farmers—except those within marketing boards—fall within one of the few segments left of the unprotected.

This inequity forces them to buy their needs at *retail* prices with built-in protectionism—protected tariffs and wages—and sell their product *wholesale,* on an international market without protection. Within this system, no route is available to them to pass their costs on.

This situation is not new; the system has always worked this way. What has changed is the protection of other segments of the production line. The farmer/rancher remains one of the few renegades of the free enterprise system.

In addition to this inequity, there is the encompassing subject of taxes. If the earlier rancher was free of significant tax, the present rancher is not. Even though few ranchers have paid noticeable income tax these past years, the effects of rising income tax with a proportionally decreased disposable income for the buying public, has greatly reduced the purchase and consumption of higher priced beef.

Since 1949, income tax has escalated four times as fast as workers' wages, while every other consumer purchase item, except energy, dwellings and interest has stayed in line, or decreased in comparative buying power. To put it another way, in 1949 one worked a day and a half per year to pay income tax; today one must work over 40 days to pay the increased income tax. Even though food has dropped in price—in proportion to today's earnings—the consumer has not the same proportion of disposable earnings left after taxes to purchase as freely the necessities of life. If, because of their poor economic performance during this past decade, ranchers have contributed very little to income tax, they have not been free of taxes. As consistent and stupendous spenders for their ranch needs, they have indirectly contributed by way of sales tax, excise tax, and duty, up to a third of these expenditures to the federal and provincial coffers. Although many products and services, purchased by farmers are free of provincial sales taxes, purchases such as lumber, hand tools, most veterinary supplies and many machinery parts and repairs are not.

Aside from these disquieting conditions that have increasingly affected the rancher's ability to survive, the most significant change from the traditional pattern has been the component of variable and high interests. Nearly non-existent as a consideration for early ranches, today it is interest that holds the prime position as an input cost. Like an extortionist, a silent partner, it demands its share of one's efforts. In 1981, in B.C., it ran at over 27% of an established farmers' incomes. As this percentage included only interest paid to authorized lending institutions, this average could in reality have been much higher.

Inflating land equity, at least in principle, may still today carry the cost of mortgages and capital improvements, but production returns cannot carry the high interest on short-term loans for farm equipment or for the gluttonous operating loan.

One does not need a very deep or broad understanding of economics to recognize that there is very little income left for the primary agricultural producer; that there is nothing left, under present day conditions for the debt-ridden beginning farmers and ranchers of this, and succeeding generations.

What legacy of a tradition can we leave the next generation except for basic subsistence or hobby farming, or the tougher route of a dedicated full-time, off ranch job to support a small herd of cattle. Under the present conditions it is beyond expectation for anyone, other than heirs to consider a working-ownership of a ranch. Is this what our country wants? Is this what the next generation expects?

THIRTEEN

THE IRRIGATION PROJECT

Man must not dispel the ambiguity of his being, but on the contrary, accept the task of realizing it.

SIMONE DE BEAUVOIR

THE MORE EXERCISE one has in doing, developing and accomplishing, the more confidence one has to take on even bigger projects. Henry had the experience and proficiency with engineering projects; I was secure in this recognition, but prior to the ranch these projects had been separate from me. Now I had to see and agree that, within the framework of the ranch, we could handle the proposed community irrigation project for the six land owners then on the hill..What's more, I knew that we had to handle the extensive undertaking, without seriously neglecting the ranch, the cattle, or our lives together.

From our beginning here we had been aware of the need for stored irrigation water. With our McCuddy Dam project scuttled by the quality of sub-soil in 1975, we had depended for late summer irrigation water on the small Heart Pond reservoir. This had been sometimes adequate for our 100 acre hayfield, but could not handle our 80 acre additional hayfield that was cleared and started in 1977.

Again Henry began his exploring, hiking, site level readings and rough surveying, as well as his aerial photo studies in search of alternate dam and reservoir sites. This time he investigated not only our McCuddy Creek system, but in greater detail, the Baldy Creek area about two miles above the ranch.

A good dam site must store the maximum amount of water for the least cost. It must either consist of an expansive, but contained swamp area, typical of beaver dam undertakings, or a narrow-necked sunken valley. The latter was more typical in our immediate area, and the type for which Henry would search.

For ease of servicing, a site must be relatively close to the ranch and have an access road or have the potential for road construction at a reasonable cost. Ideally, the site should be located at a high enough elevation to provide the head for gravity flow at a pressure sufficient to service the needs of the irrigation system.

The final essential ingredient for dam construction is for suitable construction materials to be located within a practicable distance. Clay soils that form the core of the dam are of primary importance, with an additional need for a coarser, gravely material for the second layer, and a very coarse gravel rock for the exposed surface. This outer layer protects the dam from water erosion.

Two sites were finally decided upon. One, the Baldy Dam, would require a dam roughly 35 ft. high with a breadth of 200 ft. As it turned out, it would require about 6000 cubic yards of compacted earth fill and would hold about 244 acre feet of water, or about 66 million gallons. The site, located 375 feet higher than the hayfield elevation, would easily provide the pressure to operate our irrigation guns at 125 pounds pressure.

The second site was about 100 ft. from McCuddy Creek just to the north of our Heart Pond reservoir. This narrow ravine would require a dam 40 feet high and 200 feet wide and contain approximately 180 acre ft. of water.

The disadvantage of this site, though closer to the ranch, was that it did not have the necessary elevation to provide sufficient pressure for operating our high water pressure irrigation guns. It would require a booster pump and, combined with that, the additional electrical costs to operate it. A further shortcoming was that the site was located not on the crown land, but on private land deeded to Northwood Mills, a large local logging, and lumber company.

When we had built our pipelines in 1975, we knew that to accommodate our continued hayfield expansion, larger lines should have been laid. Without adequate funding, the short-term approach at that time was taken; therefore, both sites would require a second pipeline to convey the stored water.

Analyzing the value of the alternate sites through that winter of 1977-78, the shorter pipeline savings of McCuddy over Baldy versus the additional pumping costs of McCuddy over Baldy, we scurried about in search of funding. Another Farm Credit refinanced mortgage would assist with about half of an estimated cost of $75 000 for either system, but from whom would we obtain the balance?

Appropriately, that winter the Federal and Provincial governments announced a joint agreement to fund rural agricultural development projects such as ours. This, by way of the 5 year Agricultural Rural Development Subsidiary Agreement (ARDSA), the rural counterpart of DREE, would provide up to 89 million dollars for community irrigation, erosion and flood control, feed lots, processing plants, and range development. That spring we investigated this financing possibility.

We were informed that financing might be made available to us, but not to us individually. Instead we must form a community comprised of at least 6 farms, or would-be farm properties, who could benefit from the project. This, we were aware, would require the participation of all the families then living on the hill. The costs, however, would be shared 75% from the government and the remaining 25% from the member's, with a pro rated cost per irrigated acre as the basis for individual members' contribution.

It was easy for us to conclude that with our large hayfields, we would be paying 80% of the 25% of the member's contribution. Under those terms, what would the real benefit be to us, and our real costs for the project?

We had earlier determined and projected that we must start our dam construction no later than that fall. Every year we delayed caused us losses from unirrigated hay, and each year's delay meant obvious escalating construction costs, with our income not increasing to cope with those inflating costs. If we went for the funding and the community participation, could the wheels of bureaucracy move fast enough to accommodate that deadline? Government involvement could delay the project an additional year, thereby increasing our costs significantly. We knew from experience how much critical time could be lost by the absence of key government personnel on holidays, and how slowly government decision making proceeded. Could we afford to go it alone?

As we questioned the benefits, we met with the ARDSA Co-ordinator, with the Water Investigations Branch personnel, and with the area residents. With the small contribution required from the residents, they had very little to lose and a great deal to gain. Their funds alone could scarcely cover the costs of a simple well, let alone be adequate to construct an irrigation system. For them it was a choice between no irrigation water and the project. They jumped at the opportunity for irrigation water, envisioning their dryland, small holdings as resplendent greeneries of strawberries, asparagus, alfalfa or apple trees. Thus our direction was set; the McKinney Road Water Users Community would in due course be a cooperative association with water licenses obtained and distributed to support the small holders' irrigable acres. Those ranged from 3 acres to 20 acres. Our ranch would utilize the balance of the stored irrigation water.

In considering the irrigation project we had to recognize that a new house, if it were ever financially feasible, would have to remain in limbo until after the completion of the project. Our small house, though large enough for us, was bursting at the seams with the abundance of accumulated files, books and periodicals, animal and veterinary supplies. It was obvious that it would never carry us through the extra paper of the project. Our back hall which had somehow once worked for warming and saving newborn calves, was now congested with a dozen pairs of boots, rain gear, straw hats, felt hats, gloves and mitts, which enabled no one except a skinny to manoeuver his way through the door.

My increasing need for order, and for an occasional room space of my own, determined that I should again put on my carpenter's apron, pick up my hammer and build a 200 sq ft. office and small guest room addition. With Henry's layout of the rooms, a pitch from the ranch crew, the framing and shake roof was on by winter. It was for me, and a sometimes hired assistant, to complete the addition by early 1979.

In the interim, the multitude of investigations and consequent delays kept final irrigation project approval from appearing until the fall of that year. An engineering firm had first to be contracted by the government to provide a detailed feasibility study. These studies would analyze, not only possible dam sites, the water shed and run off patterns, but also the cost/benefit ratio, the essential component for all project approvals. Even though we had previously proved that development of the once forested land increased the forage yields from 30 lbs. to the acre to 5 tons to the acre, this was not sufficient proof of the project's viability until all construction costs were calculated.

Although several alternate sites were investigated, the engineers' study found that Henry's chosen sites at Baldy and McCuddy were the most feasible and least costly sites; that in order to accommodate the present as well as the future needs of the community, both sites should be developed. In addition, the height of our then smaller Heart Pond reservoir dam would be increased by about eight feet, and ponds on two of the other members' properties would be utilized for their extra storage. These would be filled, as the reservoirs would be filled, from spring run -off, but used exclusively by the pond owners. This proved to be cheaper, even including small pumping charges, than building higher dams, or an additional dam to accommodate the total irrigation needs.

When the engineers' estimate of costs for the project were available for study, we were perplexed to see such a small estimate for dam construction. Our own earlier 'guesstimate' had placed the costs at twice as much, and this had been made a year and a half earlier. We were pleased at the prospect of lower costs, if not somewhat skeptical, but allayed any real concerns we had by recognizing that if the dam estimates were too low, the estimates for pipelines could be as much as 40 per cent too high. Victoria engineers, in the main, were not alarmed by the estimate, thus we rested our fears on the specialists' shoulders.

However, if these estimates were to have any relevance, it would be necessary to commence by early 1979. Henry had already lost his earlier projected construction time of the earlier fall, and in so doing had lost part of our second cut of hay from lack of irrigation water. A third cut had been out of the question.

When Henry is determined, there is nothing to deter him. Using our own funds, he hired one of the area's best bulldozer operators (this the operator who had so courageously built our trenches for the original pipelines) and hired the necessary scrapers and compactors to commence the Baldy dam.

Eventually, the engineers had been able to provide the dam design and engineering; the project had been approved though funds had not been released. If we wished to proceed, however, the Water Investigations Branch insisted that a resident engineer be on the site. As no government engineer could be freed, an engineering technologist was provided by our engineering firm. His job was to approve the materials used, examine and control pitch of the slope, and test for compactions by the use of a nuclear gauge.

Scarcely had the construction commenced in late October when the rains came. These fall rains were almost an unknown occurrence

Later view of McCuddy Ranch home

McCuddy Reservoir

Western view of irrigated hayfields

Rock Creek Fair Championship trophies: Tim Duursma, Henry, Eric

in our dry belt area with an average yearly precipitation at the ranch of 14 inches. The men and equipment, fighting against the weather made some progress, but it became obvious that the dam could not be completed before the winter frosts closed the job down.

During that winter, our funding was committed by ARDSA and our costs to date were to be rebated by the government. Further to this there was to be an additional claim allowed for our existing pipelines. These were to be purchased by the community at cost and were to become one of the arteries of the system.

Through a winter of feeding and ever-expanding paper work, I managed to gather together the invoices supporting the dam construction to date and search out the earlier invoices and cheques to document the costs of the original lines.

In addition, Henry had more time to study in greater detail the engineers' estimate of costs for the project. To his horror, he discovered that the costs of the original lines had not only been left out of the estimates, but were subtracted from the estimate of costs for the project. We had planned to use the rebate from these lines as a partial payment of our own costs of the project. Not being financially able to donate the lines to the community, their purchase by the community would have to take place. At some future date we would have to reconcile the budget error amounting to $50,000. In the interim, we needed to put from our minds, not only these facts, but the gnawing dam construction under-estimate as well. We were committed to the project. Even I was learning not to panic.

With the snows receding, work was again commenced on the Baldy dam. Simultaneous with this, the nearly two mile pipeline was started. Because of the rough terrain and its high cost, steel could not be used. Instead, like the earlier lines, PVC plastic lengths would follow the earlier lines. The size, however, would be larger - 12 inches - reducing to 8 inches at the ranch. This line again would be trenched and covered.

We were again able to hire the dedicated Tommy Boldt, the layer of our original lines, to act as foreman. With the assistance of up to five helpers, they were able to manoeuver the pipes down the slopes and along the canyon of the difficult route. To reduce costs, Henry, checked by Victoria engineers, had taken on the pipeline designs. This would save many thousands of dollars in engineering fees.

Baldy Dam, as we had anticipated, was running over budget. Wet weather, equipment breakdowns and the unbudgeted costs of the engineering technologist (encamped and dining at our best motor inn in Oliver) were rapidly escalating the costs. Each day of work was costing upwards of $2,500, and if that day was not fully productive, the job would extend itself and its costs.

Henry, as part time project manager, worked incessantly to make the best buys on pipes, valves, gates and fittings. With most of the pipes and fittings, manufactured in the United States, and with our falling dollar and growing inflation, costs could only escalate.

By the fall of 1980, Baldy Dam was substantially completed and along with it, the Baldy pipeline. McCuddy Dam and the additional heights increase for the Heart Pond had been commenced in the fall but were not completed before the winter set in.

Even with the care that we had taken to reduce unnecessary spending, my winter accounting as treasurer made it clear that the project would over-run by about 50%. We had to recognize that, even had the original estimates been accurate, two years had passed since these estimates had been calculated, and during these years all costs had risen some 40 percent. This, plus the omission from the estimates of our pipeline purchase, was most of, if not the total, reason for the cost over-run.

For the community, there was to be an additional cost which was not realized when the project was commenced. Our McCuddy dam and reservoir was being built on deeded Northwood Mills property. Initially, Henry had received permission from the company to lease the twenty acres of land required. They would remove the merchantable timber, leaving us with clean up costs. As our ranch already leased the 160 acre parcel as grazing land, it seemed an easy, acceptable agreement. Unfortunately, before the agreement was signed, Northwood Mills, the tree farm license, and the deeded 160 acres were sold to Weyerhaeuser Canada Ltd.

Following lengthy negotiations, Weyerhaeuser, cautious of liability problems, decided that the community must instead purchase the acreage needed.

This would require a subdivision to create the water utility lot, and with it the delays and costs in acquiring easements and complying with subdivision rulings. The ARDSA program could not include sharing of the costs of land purchase, but they could share the legal and survey costs. The community would have to provide the total purchase price of the land.

Henry or I would have to provide the leg work for co-ordinating the subdivision. With a one lot subdivision of our own underway - to help provide us with funds for our share of the rising costs of the project - and by this time running over 200 head of cows and wintering close to 400 head, we wondered where we would find the energy or time for this additional effort. Even thinking and discussing the various aspects of the project was moving beyond us.

Although we were provided with an on-site liaison co-ordinator between ARDSA, the Victoria engineers, and the project, communicating and keeping all the various departments informed: the engineers, the co-ordinator, the workers, Weyerhaeuser, the various lawyers, the surveyor, and the often as concerned members of the community - seemed a job in itself. By this time, without a vacation for over ten years, without recreation or much relaxation, like the seasoned rancher we had begun to suffer from burn-out; we had slowed down and easily escaped to an hour of the passive experience, the stultifying effects of the other world, television.

To seek escape in a fantasy world, 'Lady Black', our elegant Persian cat, graciously reclined, purring and primping, became my alter ego. She represented for me an easeful way not taken. It was simple to see that Henry too, longingly saw himself in the exuberant energy of his Border Collie, 'Corky'.

The spring of 1981 brought with it a return to old selves, and with this, the vitality and determination to complete the project. A meeting of the community members would concur with our decision. Even with the cost overruns, we were still within the established cost/benefit ratio. Additional funds were available, and the project could be completed.

The work still to be done was to complete the McCuddy dam, as well as construct a small, nine foot, north end dam to contain the reservoir from spilling out from that lower elevation. Diversions, spillways, the additional mile and half of McCuddy pipe line, and the pumping station also had to be completed.

Without three phase electricity and a projected cost of $20,000 to bring it to the ranch, it proved cheaper, if not totally as effective, to install a phase converter for the pumping station. Coincidentally, a new type solid state converter was being readied for production by the National Research Council and Delta Electric of Sarnia. Already, this converter was being promoted as alleviating the problems of the earlier types. The use of three phase, rather than single phase electricity would reduce our purchase price of the electric pump motors. In addition, in a unique way, turbine pumps were enclosed in water-tight containers so that the supply water, already under 50 pounds pressure, would not be lost. These turbines, used in this manner, gave us a higher efficiency, than the more commonly used centrifugal pumps. It also allowed us to decrease horsepower requirements from 60 to 45 h.p. to reduce electrical costs.

That summer, many workers were hired to remove the trees from the reservoir basins. Before the dams construction, these had been limbed and fallen, and floated over the surface. It was an intensive job to gather and pull into shore the limbs, dispose of them, then the logs removed, pulled to shore, and trucked away to the local sawmill.

By the fall of 1981 all work was completed except for the pumping station. To our chagrin, the pumping system itself would carry over into 1982. The converter could not be made available, thus none of the pumping system could be completed. As it turned out, because of lack of research time on the prototype model (the research staff instead had been placed on more lucrative and demanding pulp industry development) the converter could not be made available for us within any immediate future. Thus, after all the delays, we would still need to settle for the old type rotophase converter.

Through another year the demands and unending problems of the project would be upon us. To make matters worse, we again had additional unforeseen costs.Leakage in the banks of the reservoirs of both Baldy and Heart Pond reservoir was developing. Victoria, Water Management engineers insisted that the expertise of an experienced dam engineer be obtained. Years before, we had learned of a sealing compound called Bentonite which could be used to solve the leak problems. We had done some investigations as to costs and procedures to go this route should a need arise. Herman Fellhauer, the Vancouver dam building consultant, would guide us further.

During that autumn, he spent a day and a night with us enjoying true ranch hospitality with a barbecue. When the meal was finished, he shared his appreciation for this unique opportunity of staying on a ranch by saying, "I didn't think that I could make it through that big steak but it was the best steak I have ever eaten."

Teasingly, I replied, "It's all in the breed; that was Charolais beef!"

The next day a series of test holes were dug at the leakage areas, sufficient soil samples were taken to clarify the problem, and the type and amount of Bentonite that would be needed to seal the leaks. Later we were to learn that even if we were to do the repair work the cheapest way with ranch and members' labour, the estimate for the repairs would come to over $12,000.

I never easily got used to the fact of these rising costs. My earlier uncanny ability to estimate costs of jobs became suspended somewhere back in the mid '70's. Cattle prices seemed never to move upwards. Therefore, I could never bring myself to recognize that everything else did. I could not comprehend why this bagged dirt called Bentonite (a clay product) should sell for nearly $4300 a ton, almost ten times more than oats or barley, or that we should need nearly 20 tons to seal the seemingly small leaks; why the removal of the small banks by a bulldozer, layering the Bentonite by hand, then compacting and rebuilding the slope, should account for such additional costs.

At a different time, going it alone, we might have layered the banks with earth and straw and let nature take its course, or learned to live with less stored water, but decisions here were so often out of our hands.

In the late spring of 1982, we discovered that our third dam built on McCuddy Creek had developed more serious problems. With several wet summers behind us, and a high run-off from a winter of heavy snowfall, springs had developed adjacent to the downstream toe of the engineered dam. This water infiltrated the downstream toe and along with the normal seepage through the dam, saturated this toe and resulted in severe sloughing. In addition, incorrectly chosen materials for the outside zones of the dam had become unstable, which when saturated gave no protection.

We spent some very nervous weeks while engineers and consultants examined the extent and seriousness of the problem. One day at least we spent wondering if we would have to evacuate the headquarters of the ranch, including the house, as there was no way during high creek water that the reservoir could be drained fast enough should the dam's collapse be imminent. We were the only development built in its path. That was scarcely a consolation.

In analyzing our problem with these dams we hope that the government will recognize that more than just a professional engineering status is required to design and build them; that an experienced geological engineer is necessary or essential for consultation. A structural engineer may do a fine job of building subdivision roads and sewer lines but this knowledge and awareness of critical subsurface conditions may not be adequate to identify potential dam problems and how to avoid them.

In thought, we now wonder if the project will ever be a completed project. Can our debt load and cost flow requirements absorb the ever increasing cost, now running upwards of $85,000 much of it borrowed at prime plus one interest, compared with our own original $45,000 estimated cost? Will the project that was meant to help us become, in some ironical way, a main reason for our loss of the ranch? Would we have done better had we done it alone?

At times, we cannot help but be angered at the unaccountability of professional engineers as witnessed through this project. Costly errors in design, specifications, and construction occurred with no easy recourse to recover the debilitating costs of these mistakes. Perhaps it was, after all, too much too soon. But in inflationary times, and within five year government funded agreements, there can be no stages. No time between inspiration and execution, between start and finish. Every step has to be taken too fast; the cheque writing, too rushed for me even to enjoy for one brief minute the sense, in all that spending, of having been a giddy consumer, a joyful spender.

FOURTEEN

THE FUTURE

For most of us, this is the aim
Never here to be realized
Who are only defeated
Because we have gone on trying

T. S. ELIOT.

WHEN ONE HAS created something from nearly nothing, as we have on this land, it becomes nearly impossible to let go.

In spite of the fact that every step that we have taken here has been fraught with difficulties, and that government delays and endless appeals have taxed our spirits, we still look kindly on what we are doing; we still harbour a deep belief in the initiative of human beings to search out and develop opportunities still available.

We have managed to develop, in our part of the province, something nearly lost in a ranch, a total management unit completely surrounded by spring, summer, and fall ranges with no need for trucking to ranges. This in itself has made this a special undertaking, the re-building of a natural, functional unit from this semi-arable slope of B.C. We cannot now imagine selling it to another, chancing that it could be altered, neglected, or divided by another. Through this long process it has become ours.

We know how ungainly and inefficient hill farming can be, and how another could view these hills more monetarily as home sites. We too curse these hills; we know that their lyrical, but treacherous contours could better service us as pastures, but where then our hayfields?

When travelling, we cannot resist ogling the flat bottom-land, alfalfa meadows, and are only consoled by knowing that we need not re-seed our hills as frequently as they must re-seed their grass-invading valley lands. But this, at times, is only a small consolation for the broken teeth of mowers, the prematurely worn bearings, the strained and early worn tires, and the risk of rolling a tractor on hillside farms. This is ever with us.

Yet we cannot imagine giving up our generations of cow families, and our bewitching ranch, any more than we can imagine giving up our total living environment of fine neighbours, mellowed mountains, pine-scented forest trails with endless picnic sites yet unexplored, let alone give up our captivation with the ease of our charming, small town business centre. We sense in our proximity to all our physical requirements, something totally human, something once seen and known in the farming communities of England and Europe in the last century. Cities in contrast, in their alienation, their sizes and crowded inhumanity have lost their attractiveness.

I admit that sometimes I am about ready for a fabricated escape house complete with plastic geraniums, but after this kind of immersion, this depth of dedication, where does one go, what does one do to be alive with the same fire of accomplishment? We no longer know how to settle for the ordinary day to day existence. We only know how to rise to the challenges, to continue to give of our best to keep this ranch in existence.

Our alternatives are many. Financially, it has become so easy to quit and to choose sitting-back, investment returns which outperform work returns in any equity extensive enterprise.

We know that this ranch is still not on a sound financial basis. How could it be? We have, however, passed through that period which comes to nearly everyone who starts out to accomplish something. That quitting period, the giving up point, is now behind us; it is not easy for it to haunt us again. Like a marriage today, you have to want to stay together. Like a ranch during these times of high interest with little financial return, in order to avoid bankruptcy or a forced sale, you have to want to stay with it. That certainty goes a long way to save a ranch, as it does to save a relationship.

We did not choose the easiest decade to accomplish this development, and can't help but question: what if we had started this development in 1960 when gas and diesel was that acceptable 35 cents a gallon?. And within that decade, inflation saw an increase of only 24%, with 500 lb. steers bringing in a 60% increase. Whereas the 1970s saw inflation up 300% with steers bringing in an increase of only about 200%. (Western Farm Input Index).

Yet in future years, there may be even more turbulent and frustrating decades for those who still wish to pursue something with their own resources. At present, opportunities for these kinds of enterprises are diminishing, which leaves the young with chances of success and excellence only within the field of sports and Olympic competition. There they receive attention and a blessing. But for others, those who wish to operate from within a different sphere, will their objectives, dreams and enterprises be denied them?

It does seem apparent that without broad opportunities for performance, without personal choices and decisions within those opportunities, the challenge of life seems to be lost or unfulfilled.

FIFTEEN

POSTSCRIPT—HOLDING ON

Time Heals Everything, the patient is no longer there.

T. S. ELIOT

A GREAT DREAM HAD been very much a part of those first ten years. Unconsciously, I believe we had been driven by a myth of man against all odds, a little like Goliath against the giants. The struggle and the dedication within many, within us, we turned to. One that so many pioneers, often without choice had gone through to reach manhood, to reach the strength in womanhood.

But the reality of our world—and where, we found ourselves within it—soon had to be faced. For the time between the mid 70s and mid '80's would be remembered by farmers and ranchers, sole proprietors, and home purchasers as being as financially devastating as the Great '30s Depression.

Nevertheless, during those early years of the '80s, we continued to ignore or to creatively find a way to meet any financial threat we had to face. I, with my roles of road runner, ranch beautifier, builder and supervisor of two cottage additions, as well as phone call maker, letter writer, keeper of the books, harmonizer, bank dealer, was always the first to know when the squeeze was on.

Once more, there seemed to be a relative to turn to, to hold trust, through use of their dollars, for a short term emergency. In return, their loan was paid back with added high interest rate increases. For it was a fact, those who had savings, benefitted enormously from the high interest rates of the mid '70s and early '80s. As an alternative, we fortunately did have subdivision potential, and lot sales to turn to. Against our wishes, we did have to turn to those sales more and more, if only to stay one step ahead of the bankers. Here, our consolation was that by offering these properties, other families could enjoy a place of clean air nature, gardens, horses to ride the numerous back roads, a home and a few animals if they wished.

Except for those ever needed ranch and corral convenience additions, our basic ranch developments had been completed. But with cattle prices still flat, and the banks, out-of- the-blue pulling in their once generous reins, and denying further operating loans, we had no choice but to suspend further capital ranch improvements. Henry, who constantly needed a place to reach for excellence, turned his focus then in earnest on the cattle.

Using more and more artificial insemination from the top show and performance bulls, improvements in our cattle breeding program became very evident. Our offspring became more predictable in performance and conformation, nearly always predictable in temperment, which sometimes seemed ignored by breeders. Because quiet, easy handling cattle had a top priority for us, it was not unusual for us to ship to commercial auctions, even well performing females and males, who did not fit our easy handling prescription.

We, as Mann Polled Charolais guaranteed the fertility of our purebreds. We worked towards a guarantee of their temperments as well.

By the early '80s we had a crown range set-up which gave us close to a dozen spring breeding pastures. These pastures were essential to a purebred breeding program which allowed us to place 25 females with one bull for the six week breeding program. If the females were bred later on the multi bull summer range, then a sample of their blood had to be drawn and sent off for sire identification. At the Ottawa lab, all our sires were on record, with an offspring's sire quickly identified. (Was this a pre-run to DNA human testing?). We had more than 50,000 acres of higher summer range, with more than adequate grass and water, and a fall range which was adjacent to the home corrals. When snow began to fall in the higher hills, the herd, with little round up required, learned to work itself back down to these lower range gates. To reduce chances of overgrazing range grasses, our riders moved salt blocks, did fence repairs, and with movements of cattle to better grazing areas, this became the necessary part of our summers and falls.

These rough ranges gave our cattle a full test in their strength of legs, and developed ease and flexibility of their joints. Ranging of purebreds was not usual, as most breeders raise their purebreds on flat pastures. The obvious reason for pastures is to protect them, and be available to deal with their physical problems. Their higher value over commercial animals, and their potential loss on a large open range, discouraged many purebred breeders from venturing into this territory. Our range losses, however were minimal, except for one dry autumn when we lost six females. We were not astute enough, and a group of these females reached the back of a dried up pond, and very quickly poisoned themselves on water hemlock, which under normal climatic conditions was unreachable.

Bulls that were range reared did pay off for us as more and more commercial ranchers, who were used to similar rough ranges, flocked to our bull sales to buy our range tested bulls. But attracting their attention did not just happen over night.The show circuit became a necessity, and Henry, at least, was on the road at cattle shows from August until October. From 1984 onward our purebred bull sale took place in March, our female sale in October.

Earlier from spring on there was the animal selection, the halter breaking and grooming and the time needed to do this. As we were by now feeling our age, with no significant holiday time for close to 15 years, with our son, except for summers, away at college, we had to turn more, to hired help to handle these additional tasks. The auction cattle sales brought in higher, and higher dollars, but this slipped away with 20 plus percent interest rates on debts of hundreds of thousands of dollars. Our additional higher worker wages, soon totalling over 50 thousand a year left us no personal wages or investment returns. Like many other beginning farmers and ranchers, our own small wages seemed to be taken directly from drawing against our once available next year's operating loans, which left us very low contributions to our retiring Canada Pension Plan.You can understand the ranchers referring to their lot as *'to-morrow. next year'* country. Tomorrow, next year will be better, but within this time frame, it was not. Very soon, we realized the great risk of keeping bulls healthy and well for two years until their sale date. Confining a group of yearling bulls during breeding season required a large, strong compound to resist their determined need to jump the fence and chase after a female in heat. Although a two year old bull brought better dollars than we knew a yearling bull would bring, we decided to test the waters, and see if ranchers would consider paying an acceptable amount to us for a yearling bull. For them, if this worked, they could receive an extra year of breeding from the bull. With extra wintering care and feeding, this bull could continue to grow to his optimum two year size, rather than our continuing to take the risk in confining such a large number of young eager bulls.

Our first spring bull sale in 1984 was filled with mostly, eager, understanding buyers, who saw that they could get that extra year from a bull, with their added advantage of paying only perhaps 70 per cent of a two year old's price. Our bulls were all semen tested, which gave them a rating to base their decisions upon. We never went back from this decision. We had the income sooner, and fewer animals to feed and care for. It was a good choice.

Always, however, there was that yearly decision, to expand the cow herd with hope for more dollars for offspring when prices increased, or to reduce the numbers by dumping through the commercial market high priced purebred calves or yearlings to make our annual payments. Henry hated to sell well reared and selected purebreds, potentially winning animals for a quick dollar. Often he chose not to do it, when the next year proved that the purebred market was not there for the well presented females. As it stood, a yearling female cost almost as much to rear as a male, but returned at auction sale, on average only 40% of bull value. Purebred females went to that reduced number of other purebred breeders, who wanted to expand or wanted to add different genetics to their herds. Bulls had that greater market of large commercial ranchers, who needed to increase their calf weight gain by the use of higher performing bulls. For us when the beef sale market was good, the rancher could make money, and at auction could afford to raise his bid to a higher figure, to take home that bull or two he clearly had eyed.

Steering young bulls was now rarely done. We didn't want or need *prairie oysters* around our working corrals. We gave every young bull a chance to prove himself. If his performance did not prove out, he was sold instead to the beef market as a lean and fiesty yearling bull. He brought fewer dollars than if he had been steered, but he would suit many palates with his very lean meat. And I felt comfortable that we chose not to tamper with his youthful male bullish ways.

By the mid '80s a new emotional element, fear, began to enter our life. We were no longer a part of the early 1970's when everything seemed possible. A time when dynamic alternatives to do a life, to provide a living seemed endless. A new game had entered our economy, and we were a part of it, but had little or no control over it.

The strong possibility of losing our ranch became a gnawing awareness that had to be faced, at least by me. Henry preferred to focus on our purebred cattle awards for the near future. And, we were to win many championship awards during the next ten years.

As a skier Henry had learned how to win. As a rancher and cattle breeder he strove forward in the same positive direction. And as often happens under stress and tiredness, we could not share our differences, nor seemed able, anymore, to be there for each other. Our sharing, became more and more, only that of the cattle and the ranch, unable, to be there for each other as humans with personal feelings and ideas to share.

To rise out of the mire of downward thoughts, I turned to what I had put aside for nearly 15 years, the creative and spiritual side of myself. A money making way to utilize a portion of my energy was to begin to do agricultural articles. This in a small way helped the cash flow as through to the mid 1980s, no one seemed to have the courage, even though interest rates began slowly to drop, to purchase our subdivided lots to build a rural home life. Our mortgage, and some of our bank debts fell into the overdue side of our ledger. Except to throw away years of breeding, sell through the beef commercial auction market a larger part of our herd, I saw our debts, with interest on overdue interest, begin to swell at an alarming rate.

Henry found a distraction in the autumn of 1986 when the Ministry of Highways decided that they should move the old road, which ran through the ranch, into the centre of our hayfield. Their plan was to build this crossing as part of a road extension through to the Baldy Ski hill some eight miles above us. With our battle with B.C. Hydro earlier, we had some experience and feelings of success with this type of battleground.

To deal with this problem, Henry hired a surveyor to work with him to find an easy route around the ranch. The road was needed for the B.C. Winter Games to be staged at the ski hill above. We did win on this as well, but only because of a short construction timeframe. This gave Highways no time to legally expropriate the Agricultural Land Reserve hayfields which they wished to cross.

However, following our exhausting appearances at over two years of Farm Debt review hearings, our future came to a grinding unexpected halt late in 1988, when we were foreclosed on by Farm Credit Corporation. After some long discussions, they did allow us to lease the ranch yearly, with no prescribed timeframe of how long this might remain in place.

I accepted a full time job in agricultural journalism, which required that I travel a great deal to cover agricultural events. My attempt to handle my ranch and home jobs soon became an exhausting undertaking. My wish was that Henry could let go, take advantage of the significant equity in cattle, equipment and land, we would still have, and move into other talents and directions in our lives. However failure could not be an option for Henry.

In the fall of 1990, when the ranch was soon to be placed on the market by Farm Credit Credit Corp., Henry found a way to buy the ranch back. Our economic situation had not improved. However a way was found for the down payment needed to successfully repurchase our ranch. We then recognized that we must sell off a good portion of our land, beyond the home ranch, to pay out the new mortgage. This, he was successful in doing.

Nevertheless, our split in our choice of directions for our lives alienated us even further, and early in the 1990s Henry and I separated. I continued on as an agricultural journalist, became more actively involved with the farm women's network movement, editing their newsletters, doing research and attending, as an active participant, more and more meetings, conferences, and round tables on farm, health and economic issues.

Henry continued to operate the ranch at close to a loss situation until Jan. 1997 when it was sold to an Alberta rancher. The new owner was able to continue to lease the adjacent block of ranch land, which had been sold in 1991 to an absentee owner. With the adjacent crown ranges, the ranch is still a complete functioning entity. This gladdens my heart.

Continuing over these years, we have seen a continual downslide of farmers, with numbers now showing only about two percent of the population classified as farmers.

Ranchers have suffered through the devastating effects of the BSE crisis and the closure of the US border, the health undermining of beef, which turned many from beef, and the loss of most of the B.C. southern interior ranches. Those larger ranches which remain, now seem to be more commonly owned by detached absentee business owners or professionals with a ranch manager to handle their operations. A challenging advantage for the new owner, is their grand scenic land to play upon or ideally explore in a saddle, with their significant, financial advantages of farm lands' exemption from capital gains.. As well, many ranches have been purchased for return to natural wildlands by the Land Conservancy, or by Nature Trust.

In reflection for us; one could wish that our life could have had the chance for more lightness, fun and an occasional ease of living. We had certainly believed that our focused and dedicated commitment should eventually place us there. It did not.

Would we have had a better life had we stayed in the city? Henry, to continue to do battle for construction of his gifted architectural designs, I to complete my Masters program and seek work in a college. Our son to be raised in the city, within the consequential peer pressures, which he was able to avoid at the ranch; perhaps to become an archeologist that he had seen as his future when young.

We likely could have taken away a better retirement, good pension plans, and a Vancouver home to back up this retirement. But what would we have missed? A place in the sun, the beauty of healthy, and caring rural lives around us. The years of fighting against the odds for survival. The test of ourselves against the elements and economy of our times.

Henry has chosen, and was able to reopen his architectural practice, works now mainly on his own, and obtains enough home designs to meet many of his needs. I have continued my work in the areas of farm women, farm debt appeal board, agricultural writing, mentor to gifted writers at the high school, and as a volunteer in Hospice work. Eric became a Management Accountant, works for a fine company, and with his wife and children live in our valley's largest city.

Henry and I still live on the hill, in our separate homes, on small ranch acreages. From his home, down to the pastures below, he can still see the young spring calves at play. Other views, for him, back to our all engulfing past life are not known. But is the sight of calves at play... not in itself, a joy in the lingering light of spring?...

ISBN 142511976-X

9 781425 119768